French knights, led by the Oriflamme, charge down on the English division of the Prince of Wales, coats of arms drawn by Bill L... (photo by Brian Monaghan).

PSL GUIDE TO WARGAMING

PSL GUIDE TO WARGAMING

Compiled and edited by
Bruce Quarrie

[PSL] **Patrick Stephens, Cambridge**

© Patrick Stephens Limited 1980

All rights reserved. No part of this publication may be reproduced, stored in a retrieval system, or transmitted, in any form or by any means, electronic, mechanical, photocopying, recording or otherwise, without prior permission in writing from Patrick Stephens Limited.

First published—1980.

British Library Cataloguing in Publication Data

PSL guide to wargaming.
  1. War games
  I. Quarrie, Bruce II. Guide to wargaming
  793'.9   U310

ISBN 0 85059 413 8

To the memory of John Sandars, whose enthusiasm and skill in modelling and wargaming remain an example to us all.

Text photoset in 10 on 11 pt English Times by Manuset Limited, Baldock, Herts. Printed in Great Britain on 100 gsm Fineblade cartridge, and bound by, Weatherby Woolnough Limited, Wellingborough, Northants, for the publishers, Patrick Stephens Limited, Bar Hill, Cambridge, CB3 8EL, England.

Contents

Introduction, by Bruce Quarrie
Origins of wargaming ... nature of wargaming ... scales ... simultaneous movement ... painting and decorating ... frontages ... figure ratios ... how it works ... the random factor ... wargames tables ... morale ... gamesmanship 7

Ancient warfare, by Phil Barker
The historical period covered ... the character of Ancient warfare ... the history of Ancient wargaming ... troop types ... weapons ... morale and training ... points values ... choosing your army ... some possible armies ... special hints on recruiting, painting and basing 20
Rules for Ancient wargaming 35

Mediaeval warfare, by Ian Heath
The historical period covered ... feudalism ... 'knights in armour' ... tactics ... the Crusades ... arms and armour ... men-at-arms ... firearms 43
Rules for mediaeval wargaming 52

Renaissance and 17th century warfare, by George Gush
The scope of the period ... European development ... the arquebus ... the Italian wars ... Spain ... Gustavus Adolphus ... developments outside Europe ... the Ottoman Turks ... Tudor England ... the English Civil Wars ... costume ... the charm of the period 53
Rules for 'pike and shot' wargaming 68

18th century warfare, by Stuart Asquith
The nature of 18th century warfare ... the campaigns ... organisation ... weapons ... tactics 69
Rules for 18th century wargaming 82

Napoleonic warfare, by Bruce Quarrie
The Napoleonic wars ... why Napoleonic? ... weapons and tactics ... formations and strategy 83
Rules for Napoleonic wargaming 94

The Crimean War, by Don Fosten
Campaigns 1820-1850 . . . Balaclava . . . military reforms . . . weapons . . .
uniforms							95
Rules for mid-19th century wargaming			97

The American Civil and Franco-Prussian Wars, by Terence Wise
The American Civil War . . . organisation . . . weapons . . . tactics . . . the
Franco-Prussian War . . . organisation . . . weapons . . . tactics		98
Rules for late 19th century wargaming			111

Colonial warfare, by Ted Herbert
The campaigns . . . terrain . . . the problems of the period . . . and ways of
overcoming them . . . weapons . . . setting up a wargame . . . skirmish
games							112
Rules for Colonial skirmishes				125

World War 1, by Stuart Asquith
Weapons . . . the men . . . naval warfare . . . air power		129
Rules for 'No man's land' skirmishes			136

Armoured warfare 1939-45, by Bruce Quarrie
The inter-war period . . . 1939-40 Blitzkrieg . . . 1941 . . . 1942 . . . 1943 . . .
1944-45 . . . tank battles in miniature			137
Rules for armoured wargaming				148

Appendices
Bibliography						150
Model figure manufacturers and suppliers			151

Chapter 1

Introduction
by Bruce Quarrie

The origins of wargaming are lost in antiquity, but date back at least 5,000 years to the Chinese game of encirclement which is still popular under the name 'GO'. Throughout history, enlightened military commanders—such as Alexander the Great, for example—are known to have used crude relief maps and markers to plan their operations, albeit on a sporadic and unsystematic basis. Other games such as Chess or even Christopher Weikhmann's 'King's Game', invented in 1644, bear little resemblance to actual warfare and cannot be counted as wargames, although they are intellectually stimulating and today's good wargamer is often also a worthy Chess opponent*.

The evolution of the wargame as we know it today really began in the late 18th century, and board games such as those produced by SPI and Avalon Hill owe their existence to Helwig, Master of Pages to the Duke of Brunswick. He devised a game played on a gridded map with, for the first time, counters representing different troop types, each of which had a realistic movement rate. Then, in 1811, while the Napoleonic Wars were still in full swing, a game invented by Herr von Reisswitz was introduced to the Prussian court. This again used counters to represent units but broke free from the tyranny of a grid system and was played on a sculpted plaster relief model. The game rapidly spread throughout the Prussian military hierarchy as it was quickly recognised as an ideal way of training junior officers in tactical appreciation, and soon became popular in Russia and Turkey too. Thus was the *Kriegspiel* born.

Board games are, however, still extremely popular, particularly in the United States; and although their scope lies outside the confines of this volume, a brief description will direct the interested reader in the right direction. They rely upon maps, nowadays overlaid with a hexagonal grid, upon which the various terrain features are marked. Movement rates are in terms of number of hexes, depending upon the ground. Military units are represented by cardboard counters, each of which is normally printed with symbols and numbers indicating arm of service (eg, infantry, cavalry, tanks, artillery, etc); movement rate in number of hexes; offensive and defensive capabilities; and other factors which vary according to the type and complexity of the game.

The counters are set up upon the playing map and moved according to the printed factors and any delaying or enhancing terrain factors; when opposing

---

*For further details of the evolution of the wargame see Andrew Wilson's *War Gaming* (Pelican, 1970). An abbreviated version of the same material appears in David Nash's *Wargames* (Hamlyn, 1974).

counters come into contact a battle takes place, the primary consideration being the opposing attack and defence factors of the counters, plus a dice throw, which is then resolved against a combat results table in the playing rules.

The major advantage of board wargames is that they allow enormous strategic manoeuvres to be performed, on a global basis if necessary. However, they lack the visual appeal of the true wargame played with miniature figures which forms the subject of this book. Moreover, in board games, units have a tendency to be annihilated to the last man—a rare historical occurrence.

Throughout the 19th century Reisswitz's game was constantly modified by its exponents, who soon broke into two schools of thought. One retained the original flavour of the game, which was played to a rigid set of rules and used dice to introduce the element of chance always present in war; the other dispelled with these and left adjudication on the result of engagements to an umpire. In modern wargaming both systems are commonly employed simultaneously, the umpire pronouncing on situations not covered clearly by the rules, and preventing cheating!

The *Kriegspiel* progressed in sophistication after the Austro-Prussian and Franco-Prussian Wars of 1866 and 1870 when, for the first time, proper battle analysis gave the players accurate statistics on which to base their probability tables. Copies of the rules were translated into English and avidly taken up in Britain and America by such military reformers and theoreticians as Wilkinson and Livermore. As an adjunct to military planning, however, the system was still in its infancy, although the German Schlieffen Plan of 1914 and the initial British response to it had both been extensively play-tested.

A retrograde step in many respects was the publication in 1913 of H.G. Wells' book *Little Wars**, described as 'a game for boys from twelve years to one hundred and fifty and for that more intelligent sort of girl who likes boys' games and books.' Chauvinism apart, Wells' book *was* a retrograde step in that it reduced the *Kriegspiel* to a game pure and simple, in which casualties were determined by actually firing projectiles from model cannon to knock over one's opponent's soldiers. I can imagine few modern wargamers wishing to hazard their lovingly and laboriously painted figures in this manner!

Modern wargaming

Wargaming in the modern sense owes its widespread popularity to one man; a man who has come in for a great deal of criticism but whose contribution to the expansion of the hobby is unsurpassed: Donald Featherstone. In 1962 an enterprising publisher, Stanley Paul, took a calculated gamble in a totally unknown subject area and brought out Don's book *War Games: Battles and Manoeuvres with Model Soldiers*. Although not an overnight best-seller, this single book has almost certainly introduced more people to the hobby—myself included—than all of its rivals, imitators and successors combined. Although now rather dated, since wargaming has increased in sophistication as well as popularity over the last two decades, it is still an extremely enjoyable and thought-provoking book, well worth buying. Don himself is, of course, still extremely active in the field, as can be seen from the list of his other publications in the bibliography!

The other major contributory factor in the growth of modern wargaming was

---

*Reprinted in 1970 by Arms and Armour Press.

Introduction

the introduction by Airfix Products of an expanding range of inexpensive polythene figures to '00/HO' scale (approximately 20 mm). Although the majority of serious wargamers sooner or later progress to the more expensive but much more visually attractive white metal castings produced by a variety of manufacturers, it is undeniable that most youngsters cut their wargaming teeth on these plastic figures; and the Airfix trophy is a coveted prize at the National Wargames Convention each year.

The nature of wargaming

But what exactly IS wargaming? This plaintive cry can be heard at exhibitions and clubs throughout the country every week but, unfortunately, the majority of established wargamers are usually too involved with their games to take the trouble to explain properly. That is one of the main reasons for this book's publication.

A wargame is an attempt, using counters on a map or model soldiers on a sculpted terrain, to reproduce, without actual bloodshed, the warfare of a given historical epoch. The basis of any wargame is a set of playing rules. The very word 'rules' often puts off potential wargamers, unfortunately, and for good reason. It not only suggests discipline and rigidity, but commercially published rules frequently run to dozens of pages of close typescript which can be, to say the least, rather daunting. Now let us put 'rules' in perspective.

Any set of wargame rules is no more and no less than an individual's (or a group's) attempt to structure the warfare of a particular period into a form which can be used as a model in order to simulate the warfare of that period using figures on a miniature terrain.

Scales

Scale is the first determining factor. A map is a representation, to scale, of a particular country or geographical area. On a one inch to one mile map, for instance, one inch represents 1,760 yards (or, to put it another way, the scale is 1:63,360, there being 63,360 inches in a mile).

In the same way, the writer of a set of wargames rules will take into account the size, composition and rate of movement of troop units in his period, relate these to the size of wargames table he has available, and determine a wargames ground scale. This in turn determines movement capabilities and weapon ranges. For example, one might settle on one inch representing ten yards; or one millimetre representing one metre. A musket with a real-life range of around 200 yards would thus have a tabletop range of either 20 inches or about 200 mm.

Similarly, if one knows from historical records that the march rate of an average infantryman was, say, 100 yards per minute, this could equate on the wargames table to a Move of ten inches, or 100 mm, per minute. I put the word Move in capitals because the time scale may vary from one set of rules to another and from one historical period to another. In an era where the fastest-moving units are horsed cavalry, for example, one can utilise a longer game move than in a period where armoured cars are zooming about at 60 mph. Thus in any set of playing rules the game Move is determined largely by the speed at which units can physically travel, and the distances to which weapons can fire.

To make this a little clearer, consider the following comparison. In the pre-combustion engine period, the fastest troops are the mounted cavalry who can, let us say for the sake of argument, move 200 yards per minute. Prior to the invention of the rifle, the weapon with the longest range (excluding artillery) was the bow or crossbow, at 200 to 300 yards. Thus, for early historical games, a ground scale of one inch to ten yards and a game Move of two to five minutes' duration is quite acceptable. When one comes to consider tanks moving at 30 mph (880 yards per minute) armed with guns firing accurately to well over a mile, it becomes clear that a smaller ground and a shorter time scale are both required, or the tanks will not only be able to traverse the whole table in a single Move, but also shoot everything in sight from the word 'go'.

For these reasons, unlike Chess where every piece has its appropriate move, in wargames rules the ground and time scales vary widely.

Simultaneous movement

Once the author of a set of wargames rules has established his ground and time scales, his next problem is whether to make movement alternate or simultaneous. Simultaneous movement is a concept which baffles a large number of newcomers to wargaming as, in virtually every other game from Ludo to Chess, the players move alternately, and this pattern is ingrained in the subconscious. In real warfare, however, generals do not obligingly allow their opponents to move and fire first before moving and firing themselves and, as a result, many wargames operate on a simultaneous Move basis, although there have been many ingenious attempts to make alternate movement work realistically. Three out of four Wargames Research Group rules use the latter system, and it certainly has its advantages.

The big problem with simultaneous movement is that it necessitates writing orders. In other words, at the start of each game Move, the opposing players must write down brief instructions for each unit under their control. The actual physical movement of the figures on the table can then proceed at each player's own pace, since both have committed their troops to a course of action which cannot usually be altered without a change of orders at the beginning of the following Move. (Conditional orders, such as 'stay put but counter-charge if charged' or 'hold fire unless fired upon', are acceptable under some sets of rules but far from all.)

Similarly, when firing guns at each other, casualties are only removed at or towards the *end* of a Move. Thus if gun 'A' fires at gun 'B' and inflicts two casualties, gun 'B' fires back at full effectiveness before the two figures are removed from play.

Under certain circumstances, especially if an umpire is present, it is not always necessary to write individual orders for each unit at the beginning of every Move. A single blanket order such as 'advance and occupy farmhouse 300 yards to your front' may suffice, that particular unit then having to continue to try to comply with the order regardless of other events until a messenger reaches it from the Commander-in-Chief. Historically, junior officers in most armies have tended to lack initiative, either as a result of poor education or because of rigid military discipline, and the most ludicrous orders have been obeyed, even in the face of heavy casualties, as a result. The famous charge of the Light Brigade at Balaclava is a classic example.

Once the wargame rules writer has established his ground and time scales, and whether movement is to be alternate or simultaneous, everything else falls together like a jigsaw puzzle after the border has been completed. March rates, weapon ranges, rates of fire and degree of accuracy against different types of

Introduction 11

target all have to be researched from historical records, of course; and for many people this research in itself eventually becomes more absorbing than the wargame.

Painting and decorating

The same applies with uniforms and flags. Although the modern wargamer is fortunate in that he is well provided with literally hundreds of readily available books on these subjects, as well as with model figures accurate in practically every detail, the quest for greater and greater authenticity continues. Which brings me on to the subject of painting.

Wargamers fall into three categories: a minority who do not give a damn about appearance and would quite happily bring unpainted figures to the table if their opponents would let them; those who care for appearance but do not have the time or the aptitude for detailed research and painting, and who settle for an approximately 'right' appearance when all their figures are deployed *en masse;* and another minority for whom appearance is all-important, who paint well and enjoy it, and who like nothing better than to spend an entire Saturday in the reference library researching the most minute details of buttons and lace.

The vast majority of wargamers fall into the middle group, and it is worth considering where YOU would place yourself before even contemplating which figures to buy.

For many years the only figures available to wargamers were those generically known as '54 mm', this being an arbitrary but standard height initially adopted by Messrs Britains Ltd and subsequently taken up by most other manufacturers. Then there was '30 mm', a scale based on that used by 'flat' figure manufacturers on the Continent. This is the scale used in Brigadier Peter Young's delightful book *Charge!,* now sadly out of print.

Both of these scales have the inherent drawback that the figures are fairly large, so that even a company of infantrymen (let alone a regiment or division) occupies a lot of space on the wargames table. They are also expensive, and require more detailed painting.

Thus the '20 mm' figure came into vogue. Ideally suited to wargaming, it remained popular throughout the 1960s and has only fairly recently been superseded by the '25 mm' figure, a compromise allowing detailed painting at economical cost. Most of the figures you will encounter at wargames clubs and exhibitions are '25 mm'. Even so, you will quickly notice discrepancies between manufacturers, since not everyone is agreed on exactly what is meant by '25 mm'. Some people take it as the height of a figure from the bottom of its base to the top of its hat; some from the soles of its feet to the top of its head; and endless permutations therefrom. My advice, once you have decided which historical period you personally would like to concentrate on, is to visit one of the military modelling and wargaming shows which are held up and down the country each year (and usually advertised in *Military Modelling* magazine) and examine the different makes and ranges before settling on one.

However, the choice is a little more complicated today since several other figure scales are now available to rival '25 mm'. Of these the most serious contender for the 'wargames scale of the future' award is undoubtedly 15 mm. Nearly half the size (and price) of '25 mm' figures, these models can also be painted in half the time because the amount of detailing you can do is restricted. Moreover, because of their size, you can get far larger numbers on to a wargames table, so battles look much more realistic. (There is also a limited range of '9 mm' figures available in America, but these have not proved as popular.)

Even smaller are the '5 mm' figures (1:300 or 1:285 scale) which are ideal for World War 2 battles where speeds and weapon ranges necessitate a small ground scale. They do, however, lack 'personality' and, as far as I am concerned, are only one step removed from 'board games', played on a map with cardboard counters to represent units.

Your choice, however, as a newcomer to wargaming, is wide open. Hopefully the photographs in this book will give you some idea of what scale is likely to appeal, but I still say there is nothing better than examining the different makes at close quarters at an exhibition.

Frontages

Figure scale is, to an extent, determined also by the ground scale adopted in the set of rules you are playing to because of unit frontages.

Throughout history, armies have all adopted standard intervals at which the soldiers in the ranks stand (or sit, if mounted). These intervals are primarily determined by the amount of elbow room a man needs in order to wield his weapon efficiently, be it sword, pike, bow or musket; and also by his role in battle, either as a shock trooper or as a skirmisher.

In the Napoleonic Wars, for example, the frontage assigned to a British infantryman in the drill books of the period was 22 inches. An average battalion of 700 men, drawn up in a two-deep line, would thus have a theoretical frontage of some 214 yards. In practice it was more, because under battlefield conditions it was obviously impossible to maintain precise 22-inch intervals, and also because of the additional intervals between companies in the battalion. But you will understand the principle.

Once you come to transfer this battalion to the wargames table, you can see that it should occupy a scale 220-odd yards of front, ie, 22 inches at ten yards to the inch or 220 mm at one millimetre to a yard. It goes without saying that frontages, moves and ranges can be expressed in metres, paces or any other convenient measure—I use yards purely for the sake of example.

If you then take into account the fact that the base of an average 25 mm wargames figure is some 10 mm wide, it becomes immediately apparent that there is no possible way you can get 700 figures, even in two ranks, into a 220 mm frontage (it would, in fact be 3,500 mm).

Nor would you really want to, since building up and painting just a single battalion of 700 figures would be at least a couple of years' work for the most prolific painter, so how long would it take you to assemble an entire army?

Figure ratios

The way wargamers get around this problem is by scaling down and establishing a 'man to figure ratio'. In other words, we say 'let one wargames figure *represent* ten (or 20, 30, 40 or 50) actual soldiers'. At 20 to one, therefore, our 700-man battalion becomes reduced to 35 figures who, spaced appropriately in two ranks, would quite happily fit the required 220 mm frontage on the wargames table.

If you were using 15 instead of 25 mm figures, it should thus be clear that you can do one of two things: you can either halve the ground scale, using the same number of figures to a 110 mm frontage, with half a millimetre equalling one

Introduction 13

yard; or you could halve the man to figure ratio to ten to one, giving you 70 figures on a 220 mm frontage. In the smaller scale you thus have scope for either building up a much larger army in terms of number of units, or one which is more realistic in appearance because the individual units are themselves larger.

Scales vary widely in published sets of playing rules and, although most are currently geared to the 25 mm figure, you can see how easy it would be to adapt them for 15 mm.

Unit frontages are important in two vital aspects. Firstly, they are an accurate means of representing the actual amount of terrain a real military unit would occupy on the battlefield; and secondly they determine the number of men able to fire or fight in a mêlée (hand-to-hand encounter). This is especially critical when you come to a mêlée between dissimilar troop types armed with different weapons and thus occupying dissimilar frontages: a unit with the same number of men as its opponent but more densely packed into a shorter frontage will have a decided advantage.

How it works

Now, I can hear many readers asking: all this is very well but plastic or metal figures cannot actually *fight* each other, so how do you go about simulating it?

At the simplest level, back in the days when wargaming was in its infancy, it was very simple: you merely threw a dice for every six figures firing or fighting, and whatever you scored was the number of casualties inflicted on your opponent. Hardly realistic, is it, after all the work we have put in getting accurate ground and time scales and unit frontages? Unfortunately, this image of wargaming as a mere game of chance determined by the throw of a dice persists in the public mind, to the hobby's detriment.

What actually happens is that anyone compiling a set of rules goes back to historical battle accounts and, where they exist, scientific weapon performance experiments. The more widely read he is, the more realistic his assessments are likely to be and the more accurate his wargames rules. The only test of this is whether they produce battles in which the average casualty levels are similar to those in historic engagements of the period, and whether the results are realistic.

To put it at its simplest, it is pretty apparent that in hand-to-hand combat an armoured knight with a long sword and shield stands a better chance of killing an unarmoured peasant equipped with a scythe than *vice versa*. The fun comes in deciding his probability of success, which is what the playing rules for this period will be based upon. If one reads an account, to give an imaginary example, of ten such knights routing a rabble of fifty serfs, inflicting 20 casualties for the loss of only one of their own number, then an accurate assessment can be made and the wargames rules based on the result. However, circumstances could alter the outcome. For example, the peasants may ambush the knights, and with the element of surprise could conceivably kill more whilst losing fewer of their own men; or the ground might be boggy, which would slow the armoured knights down more than the peasants, and thus change the result.

From just this single example you can quickly see yet two further factors which the competent rules writer has to take into consideration: surprise and terrain.

Similarly with missile weapons. You might find a battle report in which, for the sake of argument, one hundred archers each fired one hundred arrows, and between them inflicted fifty casualties, giving a one half per cent probability of a hit per arrow. Unfortunately, statistics are rarely this obliging! In my own favourite Napoleonic era, for instance, three different reports on tests of smooth-bore musket accuracy, one English, one French and one German, give widely disparate results of between 53 and 75 per cent at 80 to 100 yards' range, 25 and 40 per cent at 200 yards and 16 to 23 per cent at 300 yards. All one can do in a case like this is to take the average and extrapolate from these for the intermediate ranges.

Relating this to the wargames situation, the Napoleonic rules writer would then further take into account the fact that results on a battlefield could be expected to be lower than in a test, for heat, movement, smoke, confusion, fear, excitement and the fact that the enemy is firing back would all exert their toll, resulting in a higher proportion of mis-fires and a much lower overall level of accuracy. Brigadier Hughes, in his excellent book *Firepower,* says that the average hit level in a Napoleonic battle was only three to five per cent.

Having accomplished this the writer would then be in a position to say that, under normal circumstances, if x men fire for y minutes at z range, they could expect to inflict a casualties on target type b. Target type is, of course, a determining factor, because it is much easier to inflict casualties on a densely packed body of infantrymen than, for example, on a more quickly moving and more widely dispersed unit of skirmishers.

The same principles of research and extrapolation apply in all periods. For instance, if you are considering tank warfare you would first go to one of the dozens of books which give weapon effectiveness and the thickness of different tanks' armour plate, and from this you would be able to say with confidence that gun x, firing ammunition type y at z range would be able to pierce the armour plate of a tank a with thin armour plate but would be unable to affect tank b with thicker armour. (It is worth pointing out in passing that these technical books rarely agree, hence wide disparities between different sets of rules!)

The random factor

However, luck enters war as every other sphere of activity (Napoleon himself said he tried to choose 'lucky' commanders) and, however painstaking and accurate your historical research, the results are only an *average* and this, of course, implies that they could indiscriminately be better or worse. This is where the dice come into things. However, in most modern sets of rules they are not allowed to affect results to too great an extent, or you would be back to a pure game of chance. In our Napoleonic games, for example, we use a six-sided dice on which two faces are numbered one, two are numbered two and two are numbered three. If a player throws a two, his results are taken at the basic (ie, average) level written into the rules; if he throws a one they are approximately ten per cent less than the average, whereas if he is lucky enough to throw a three, they are approximately ten per cent better.

Different rules use different systems. For example, there are 'average' and 'percentage' or 'decimal' dice. The former are six-sided but numbered two, three, three, four, four, five, three and four thus being the 'average' which gives them their name. Percentage dice are ten-sided and numbered from zero to nine. Two, usually of different colours, are thrown simultaneously, one representing tens and the other units. The result is an actual probability. Thus, for example, if a tank had a 50 per cent probability of hitting a particular target, a score

Introduction

15

between 01 and 50 would denote a hit, 51 to 99 and 00 a miss, 00 representing a hundred.

Wargames tables

Model soldiers, playing rules and dice. What else is required in order to begin wargaming? Answer: very little. An expanding steel rule marked in both inches and millimetres will be required by each player in order to measure their troops' movement and weapon ranges (an ordinary household tape measure will do in a pinch). A pad of paper and a pen or pencil will be needed to write orders and keep a record of casualties. A pocket calculator with memory store and percentage key is useful when you get on to more advanced games, but is certainly not essential.

What *is* essential, of course, is somewhere to play, and some model terrain to represent houses, trees and hills, etc.

The ideal wargames table is no more than five or six feet wide (that being the maximum over which you can reach units in the centre in order to move them). Length is indeterminate and will usually be decided by the size of the room in which you will be playing. The average table is between five and 12 feet long. These dimensions, I must add, are for games with 25 mm model soldiers. If you choose 15 mm, a table half this size would be adequate, or if you keep to the larger size it allows more room for manoeuvre and more units on the table.

Few wargamers can afford the luxury of a permanent room devoted to their hobby, and most are resigned to usurping the dining room table in between meals. However, the average such table is too small for a decent wargame. The most common way of surmounting this problem is by means of a ply or chipboard 'overlay' to the desired size, preferably with two by two cross-braces to keep it rigid. This board is placed on top of the dining room table, overlapping the edges by a couple of feet all round. Two points should be watched, however. First, ensure that the table is adequately protected against scratches by layers of newspaper or an old blanket. And secondly, when your board is in place, don't lean on the edges or you will have the whole lot over!

The board itself can either be painted green or further covered with a green baize or felt cloth, providing a basic playing field. Model trees and houses can then be placed upon this. These are available from a variety of manufacturers, the houses in card or plastic, the trees from a model railway stockist, or they can be home-made. Twigs from the garden, glued to a cardboard base with epoxy resin, and their 'branches' covered with wire wool which has been sprayed green, make excellent and inexpensive trees. The foliage effect can further be enhanced, while the paint is still wet, by sprinkling dried tea leaves over the trees, then spraying a second coat of a different shade of green. (For obvious reasons you should use matt paint, not gloss. These are available in aerosol spray cans from all good model shops.)

Similarly, rather than buying expensive proprietary products, hedges can quickly and simply be made from strips of foam rubber 'plucked' to an irregular outline, fences from matchsticks and walls from Plasticine. (Coated with Banana Oil to seal it, Plasticine can be painted and becomes quite rigid.)

Houses themselves can simply be constructed from cardboard with doors, windows and beams simply painted on or, if you have a more practical bent, can be built properly using the special brick, slate and wood textured card which, again, is available from model railway shops.

Hills pose a problem because, although they can be created realistically using a modelling compound such as Mod-Roc over wooden formers, model soldiers will be unable to stand upright on them unless the slopes are very slight. For this reason the most common type of hill seen on wargames tables is constructed from layers of polystyrene ceiling tiles, cut to irregular shapes and painted in varying shades of green, brown and yellow. These have an inevitable and unnatural 'stepped' appearance but at least provide flat surfaces on which the figures can stand.

Roads and rivers can be made most simply by drawing them in with brown and blue chalk, or by laying strips of brown and blue cloth on the table. Alternatively, a much improved appearance can be obtained through making road and river sections (like lengths of model railway track) out of cardboard, indicating the banks with Plasticine, and painting accordingly. Hedges can be permanently attached to road strips, as can trees and fences; and river lengths may be enhanced by the odd patch of 'reeds' made from the bristles of an old brush, and even perhaps the odd miniature duck or moorhen!

For most people, this is as far as wargames terrain ever goes, and it is surprising the number of wargamers who spend hours painting superbly detailed figures only to let them down by hastily and poorly constructed terrain. Much more realistic (although leading to storage problems) are terrain *modules*. Here, the industrious wargamer constructs interlocking two-foot square sections of terrain which can be pushed together in different configurations to give practically endless varieties of battlefield. Each terrain module is properly sculpted, with a gently undulating surface covered in grass flock or grooved and painted to simulate ploughed fields, etc. Hedges, streams, trees, roads and buildings are all incorporated, each module in fact being a miniature landscape diorama just waiting for the model soldiers which will bring it to life. As I said, storage of these units can be a problem, but if you have the space the end result is well worth the time and effort invested.

Planning these modules so that they will interlink takes some thought: for example, roads and rivers must both begin and end either at a corner or the exact centre of each. Much more flexible, although it does demand a permanent wargames room in the house, is the sand table.

A sand table is simply an ordinary—but strongly built—table to the required dimensions, whose top surface is enclosed by six-inch high walls. Sand is poured into the cavity thus created to about an inch below the top. Dampened and then sculpted, the sand can quickly be made to resemble any conceivable battlefield; trees and other features can simply be inserted into it, and a 'countryside' rather than 'sand dune' effect created by gently blowing powder paint in various green, brown and yellow hues over its surface. (After a while the sand becomes impregnated with these and requires little if any additional colouration.)

Sand tables are enormously flexible but are also enormously heavy. Before trying to build one you must, literally, test to see whether your floorboards and joists will accept the strain. (Sand tables, I must add, are rare among 'ordinary' wargamers, although used widely at Sandhurst—no pun intended!)

If you venture into other fields such as naval or aerial wargaming, for example, or even science fiction space battles, 'terrain' becomes less of a factor. In a naval game with model ships arrayed against each other all that is required is a greenish-blue board, perhaps tastefully picked out with white flecks representing wave crests; shoals, rocks, wrecks and other obstacles are simply

Introduction 17

shown on each player's individual chart of the area in which the battle takes place—and may be more or less accurate according to the umpire's discretion! In an aerial game played according to the rules outlined in Mike Spick's book on the subject (see page 151), a plain blue board is all that is required, clouds being represented by irregular shapes in card or polystyrene. In a sci-fi game all that is needed is a black board with planetary orbits and the gravitational influences of stellar bodies clearly marked, upon which you can manoeuvre your model space ships.

Speaking of science fiction necessitates mention of the current fad for fantasy wargaming, which is only limited by imagination but which, by concensus, is not really *war*gaming at all! It has been a truism for many years that a high proportion of wargamers are also science fiction/fantasy novel enthusiasts, and this fact has more latterly revealed itself in the reverse direction. Fantasy gaming usually takes place in a mythical world populated by dragons, orcs, elves, mighty warriors and exotic maidens, wizards, witches, hidden treasure troves and other unlikely ingredients. The most popular form—'Dungeons and Dragons'—bears similarities to the skirmish wargaming system described later in this book, in which each player controls one or two characters, each of whom is assigned different characteristics for such factors as strength, dexterity and intelligence by means of a random dice throw. These characters form roving bands of adventurers who set out to explore mysterious and magically guarded caverns whose various inmates are controlled by the umpire or 'dungeon master'. While good fun as an evening's entertainment, this type of fantasy is not really 'wargaming' and is therefore outside the scope of this book*.

Morale

We are nearing the end of this introduction as the newcomer to wargaming has only two more salient factors to consider before plunging into the ensuing historical chapters and deciding which period most appeals to him. These are morale and gamesmanship.

Just as model soldiers cannot bleed, so they cannot experience emotion, be it fear or lust. So the wargamer must inflict these human frailties upon them. Why?

Consider two regiments of infantry. Both have been recruited from a civilian population with a similar background, be it agricultural or rural, rich or impoverished. The ages and physical stature of the two groups are essentially identical. The training, discipline, food, pay and weapons are also similar.

Now put them into a battle. One regiment throws down its arms and flees, the other suffers horrendous casualties but fights to the finish. Again, why?

Essentially these are both unanswered questions, or by this time someone would have forged an unconquerable army. However, certain parameters can be partially accounted for. To give an example: if one of the above regiments has inspired leadership, occupies higher ground than its opponents, and has both of its flanks secured by either natural or man-made obstacles (so that the enemy cannot circumvent it and attack from the rear), then the spirits of its men would be higher than of those in a comparable regiment labouring under unimaginative leadership, overlooked by enemy troops on higher ground, and with one or other of its flanks menaced.

*At the time of writing, however, the publishers of this volume had in preparation a book entitled *Fantasy Wargaming*, by Bruce Galloway, which should be available in 1981.

This is an extreme example but it is upon such bases that 'morale' rules are established. The writer of a set of rules takes into consideration such factors as height, flanks, cover, casualties, strength of supporting *and* opposing units, tiredness, etc, and allocates a plus or minus factor to each. The numerical level of any given unit's 'morale factor' then determines whether it stands and fights or runs away.

This is particularly important when you look at accounts of actual battles, for it soon becomes apparent that most fighting units (and I do stress the 'most') give up the battle long before they are even reduced to half strength. There are inevitable exceptions to prove the 'rule': of units fighting to the last man; of cavalry charging straight into the mouths of serried cannon; and innumerable others. This is why such awards as the Victoria Cross were instigated. Because such actions are beyond what can normally be expected from human flesh and blood.

The compiler of a set of rules cannot, unfortunately, cater for the exceptions except by making them conditional on an exceptional dice throw. If the rules are to work at all reasonably, they must be arranged around the average—although the dice alter the percentage somewhat.

By the same token, the result of a wargame cannot be left to the willpower alone of its commanding generals (the players) as some unimaginative and ostrich-like pundits would like to have us believe. The will—or lack of same—of the troops themselves to fight *has to* be taken into account. Once you begin delving into military history, as any wargamer must sooner or later if he/she wishes to improve his/her game, this becomes immediately apparent. A commander can exhort his troops with his dying breath, but unless they believe in him, believe in their 'cause' to a greater or lesser extent, and believe in their ability to win, then the will to fight will be extinguished and the unit(s) in question will scatter to the four winds.

A good set of wargames rules takes this into account. The factors affecting morale vary considerably, not just from one epoch to the next but also between nationalities. For example, in an Ancient game morale will be boosted if the seers predict a victory from 'reading' a bull's liver; in a Napoleonic game the morale of Prussian troops in 1813-14 will be boosted by the prevailing nationalistic fervour and desire to rid their country of the French yoke, whilst the morale of their opponents will be diminished by the recent disaster in Russia.

Morale is certainly the most ambiguous concept in wargaming; the most difficult to tabulate, and the most variable. Yet without morale rules one might as well revert to H.G. Wells' system and leave victory or defeat to be determined by the accuracy of each player's matchstick-firing gun. Not all players share this attitude, however. If you encounter one you can be assured that he is a war*gamer* rather than a *war*gamer.

Gamesmanship

Which leads us logically into my final remarks on gamesmanship. Wargaming is many things to many people. For some it is a pure entertainment in which the appearance of the troops on the table is paramount; for others it is a real challenge in which victory is all-important. Of course, we all like to win—but please!: can we have victory without acrimony? The wargamer who argues bitterly the roll of every dice, consistently moves his figures a few millimetres

Introduction

more than he should, contests every arc of fire and orders 'psychic' moves which a real commander on the spot could not possibly do, is very likely to find himself without opponents in extremely short order.

So, my number one rule for all prospective wargamers: don't argue! If a situation arises which is not covered by the rules, resolve it by tossing a coin. By all means discuss it *after* the game and, if necessary, amend the rules accordingly for the next time. But don't let it interfere with the amicable progress of the battle in hand.

To give a specific example of what I mean: I recently received a letter from a young man who plays according to the rules published in my earlier book *Napoleon's Campaigns in Miniature*. Under these, the effectiveness of artillery fire is determined to a large extent by range. At 500 yards, for example, a 4 pdr gun would suffer a minus three deduction whereas at 600 yards it would be minus six. However, cannon balls do not stop dead where they land: they bounce and continue inflicting damage for some considerable distance beyond their initial impact point. In the case of a 4 pdr the rules give this 'penetration' distance as 200 yards (depending on the terrain). What my correspondent's opponent was doing, therefore, when firing at a target 600 yards away, was announcing that he was actually aiming at a point a hundred yards in front of it (thus bringing him into the minus three rather than minus six bracket) and claiming a hundred yards' penetration into the target. This is called playing to the *letter* rather than the *spirit* of the rules and is how several players have established their reputations. You can't, unfortunately, call it cheating. All you can do is avoid playing with these people again. As Phil Barker once said to me: 'if it sounds like a fiddle, you can't do it'.

In all seriousness, wargaming is a hobby whose principal aim is relaxation. Take it seriously by all means when it comes to studying uniforms, weapons and tactics, because these will broaden your knowledge of your chosen period and, inevitably, improve your tabletop performance. But don't approach a battle with a 'win at all costs' attitude. You will not only learn a lot more but will also be assured of regular opponents to play against.

To conclude, it is obviously impossible in a book such as this to answer everybody's questions. The most it can hope to do is signpost the way. For that reason we have included a select bibliography from which you can obtain further information. Virtually all of the books listed can be ordered through your local bookshop or, if finances will not stretch that far, from the library. If you still have questions, *Military Modelling* magazine (incorporating *Battle for wargamers*) runs a very efficient question and answer service with a panel of experts on uniforms and equipment. Moreover, the contributors to this book, myself included, will be pleased to help if you write, enclosing a stamped, self-addressed envelope, c/o Patrick Stephens Limited, Bar Hill, Cambridge, CB3 8EL, England.

Cambridge,
June 1979 BRUCE QUARRIE

Ancient warfare
by Phil Barker

The historical period covered
In the early days of wargaming, 'Ancients' was accepted as covering organised warfare from the very earliest times until the introduction of missile weapons deriving propulsion from the explosion of gunpowder. It is more usual today to split off the later part of this period, extend it to take in the earliest primitive gunpowder weapons, and call it 'Mediaeval'.

Where precisely the Ancient period should end and the Mediaeval period start is a matter of some doubt. The *Cambridge Ancient History* takes mediaeval times as starting with the first Christian Emperor of Rome in 306. The insular English start it with William the Conqueror in 1066, mainly because that is the only date most of them can remember! To make things more difficult, all the ingredients considered to be typical of Mediaeval warfare such as heavily armoured lancer cavalry, powerful longbows, crossbows, spear and pike phalanxes, siege catapults and feudal organisation are also, though less commonly, found in Ancient warfare, and the most popular set of wargames rules for the Ancient period covers Mediaeval as well. In order to leave Ian something to write about in his Mediaeval section, I have chosen somewhat arbitrarily to end with 650 AD.

At that time, the Sassanid Persian Empire had been conquered by Arabs inspired by a new religion, the Byzantine Empire had survived the initial Arab onslaught but at the expense of losing half its territory and having to reorganise its army on a different basis, and the barbarian conquerors of the western half of the Roman Empire were beginning to settle down into the familiar mediaeval pattern of noble mounted warrior supported by an understratum of peasants. The three main ingredients for the Crusades, Saracens, Byzantines and western knights have, therefore, almost arrived.

This cut-off point still leaves plenty for me. Lasting from approximately 3,000 BC to 650 AD, the Ancient period is still nearly three times as long as all the other wargaming periods combined. It is this that gives it the sheer variety that is its greatest charm.

The character of Ancient warfare
The main characteristic of Ancient warfare was that battles were decided by hand-to-hand fighting with sword, spear or similar weapons against enemy within arms' reach. Longer range missile weapons such as bows, slings and javelins were in use, but because of the prevalence of armour and shields could

Ancient warfare

not cause enough serious casualties to win a battle on their own. However, if efficiently utilised and not appropriately countered, they could weaken a force to the point where it could no longer resist effectively hand-to-hand.

The effective defence provided by shields and body armour also slowed the rate at which casualties were received and inflicted in hand-to-hand combat. Armour usually protected only the vital areas of the head, trunk and abdomen, however, leaving the arms and legs exposed to non-fatal, but ultimately incapacitating, wounds. When a force did break and flee, even a slight wound might be enough to ensure that a fugitive was overtaken and slain, so that the winners' casualties would have a very high proportion of lightly wounded to killed, the losers an equally high proportion of dead to wounded. This, and not prejudiced reporting, largely accounts for the great disparity between winners' and losers' losses that we find in contemporary battle accounts.

The slowness with which casualties were produced by either spear and sword or by missiles led to the invention of special methods of attack relying on the kinetic energy of impact to knock defenders off their feet and break up their formation. The impact of a galloping horse, either carrying a rider or pulling a chariot, was the most usual means, but some nations went an order of magnitude further by using elephants. Relying as it did on the velocity of the animals at the moment of impact, the method could be countered by breaking the momentum of the charge with a volley of missiles or an array of closely spaced spear points, or by depriving the charge of a solid target to hit.

War at sea was very much like that on land. Ramming was seldom effective in sinking a hostile ship unless it was relatively small, though it could often cripple by breaking oars and injuring the rowers who, incidentally, were free men and not slaves. Fire was a very two-edged weapon, and the sizes of catapult artillery that could be installed on warships were mainly effective against personnel.

Right wing of Late Roman army formed for battle. Light cavalry of the Promoti and Scutarii on the extreme right, then Catafractarii guard the outer flank of the infantry with the Alani in support behind. Two Palatine Auxilia units, the Cornuti and Victores, form the first infantry line, with the Legions of the Lanciarii and Herculiani in support. The Emperor and his bodyguard wait behind the junction of the cavalry and infantry. Figures by Miniature Figurines.

Late Roman artillery with stone-throwing heavy engines and bolt-shooting light engines deployed behind a screen provided by the Batavi. Pack animals and draught oxen wait in rear. The engines are by Hinchliffe Models.

Boarding fights were the rule rather than the exception, and the normal complement of marines was often assisted by land troops drafted on board or by armed rowers.

Most cities were fortified with stout walls and sieges were frequent. Only the largest of the mechanical stone-throwing artillery were capable of producing a breach by bombardment, and these had to be set up dangerously close to do so. Other methods of breaching included undermining and battering rams. Another route was over the top of the wall, either by scaling ladders, from a movable siege tower pushed against the defences, or from a specially constructed earth mound. It was usually necessary to establish a measure of artillery superiority over the wall or tower-mounted machines of the defenders before breaching, mound constructing or moving up of towers could get properly under way. A city that fell to assault instead of surrendering in good time was not entitled to the attackers' consideration and was sacked with concomitant atrocities.

The history of Ancient wargaming

Unless one counts the Indian and Chinese boardgames that developed into our modern game of chess, there was no historical equivalent to the 'Kriegspiel' military training game from which modern 'Horse & Musket' wargaming originated. The first set of rules that I know of were the Cass-Connett rules of around 1960, which introduced many techniques incorporated a little later by Tony Bath into the set that was to become famous due to its inclusion in Don Featherstone's bestseller *War Games* in 1962.

The only Ancient miniature soldier figures available at that time were German 30 mm flats, which were very difficult to obtain, and home-made conversions

Part of a later Byzantine army. The front rank from left to right consists of Dixon Turkish light cavalry, Hinchliffe extra heavy and super heavy cavalry, and Hinchliffe light infantry bowmen. The rear rank has Lamming Norman mercenaries, Hinchliffe Varangian Guard axemen, and Lamming heavy infantry spearmen.

from Airfix 20 mm plastics of other periods. This was blamed for the relative unpopularity of Ancients by Tony in a 14-page handbook on Ancient wargaming which he wrote for Don's *Wargamer's Newsletter,* gestetner printed on one side only, torn off with a ruler instead of guillotined, and roughly stapled together.

Today, the Ancient player can get almost any conceivable type of Ancient troops in both 25 mm and 15 mm scales and can choose between or mix the figures of several different manufacturers. When he comes to paint and organise them, he will find a range of scholarly and professionally produced books with copious illustrations to tell him everything he needs to know. He will also not be short of opponents, because Ancients is now possibly the most popular of all wargaming periods.

One of the major factors in this change was Tony Bath's founding of the Society of Ancients in 1965. The Society publishes a bi-monthly magazine, *Slingshot,* runs an international league competition, and has a world-wide membership that is far more influential than its current number of approaching 3,000 implies.

In 1966 came the first British National Wargames Convention, held initially in Southampton, but subsequently hosted by the previous year's winning club. From 1969 onwards, the rules used have been those produced by Wargames Research Group, of which I am a member.

The Wargames Research Group Ancient rules came into being through the discontent of three keen Ancient players with existing sets. There were three main reasons for this discontent. Firstly, the model figures were not behaving like real troops, combining bravery, telepathy and stupidity to an excessive degree. Secondly, the capabilities of the weapons had been based on secondary historical sources of, to say the least, dubious accuracy. Thirdly, a majority of wargamers felt that the mechanisms employed in the rules depended a little too much on dice or similar chance devices.

Getting together after the 1968 Nationals, the three of us agreed that Bob O'Brien of the Worthing Club would go away and think up a radically new and more accurate way of simulating morale and the problems of command, that Ed Smith of London would try to find new mechanisms, and that I would go back to contemporary military manuals and histories to try and establish the true characteristics of the weapons and act as editor-in-chief.

Early trials were successful enough that the Worthing Club, who were hosting the Nationals that year, adopted them for use in the Ancient competitions. All the other sets used were written by Stephen Reed of the Worthing Club and vanished without trace after the Convention. Not so the WRG Ancient rules. These were an amazing success and spread round the wargaming world like wildfire.

This success was rather embarrassing and in the end forced us to go professional. We developed other rule sets for other historical eras on the same principles and our mechanisms were very widely adopted by other rule writers, so much so that too little attempt has been made since to introduce alternative methods. We have tried to find new mechanisms ourselves, but although some of these have been very successful in other periods, they have not yet proved any great advantage in Ancients.

The WRG Ancient rules have been periodically updated and the era covered expanded. The 6th edition, which should be finalising as this appears in print, covers the period between 3000 BC and 1485 AD, incorporates the latest historical research and utilises a slightly improved set of mechanisms to speed play and reduce the need for umpiring.

A number of rival sets have been produced, but none of these have made any great impact. Most of them have retained WRG troop definitions, weapon classifications and base sizes, and it is likely that this will continue to be the case with future sets, indeed may be a prerequisite before players are willing to buy them.

The almost universal acceptance of the WRG Ancient rules has been a major factor in expanding interest in Ancient wargaming. You can now go to any English-speaking country and many others and be sure of getting a game with them. However, no one knows what the future may hold. It is possible that one of you may be inspired by this book to sit down and write a set that will prove equally popular or to invent new mechanisms that are just what we need!

Troop types

Those military manuals of the era that survive to our day provide us with a troop classification that forms the basis for that in current rules for Ancient wargames. They first distinguish between land and sea forces, then between fighting men and logistical support troops such as surgeons, camp servants and muleteers, then between those land troops that fight on foot and those that fight mounted on animals or in vehicles.

Foot soldiers are divided into three classes called in Greek Hoplites, Peltastoi and Psiloi. Because these general terms are also used for very specific soldiers within those classes, it is more convenient to refer to them as close formation, loose formation and dispersed formation infantry.

Close formation infantry fight with friends close by on either side. This maximises the force they can bring to bear on a given enemy frontage, but puts them at a disadvantage in terrain which tends to break up their formation such as woods or rough hillsides, or when crossing linear obstacles such as streams. Their primary weapons are usually intended for fighting at close quarters. These can be thrusting spears used overarm in the right hand, long pikes grasped with both hands, a long axe or similar heavy cutting weapon wielded in both hands, or a heavy spear or light axe flung immediately before contact with the enemy and so not counted as a true missile weapon. They all carry a sword or equivalent secondary weapon, and some may have light hand-hurled javelins or darts as well, giving them a limited long-range fighting capability.

Some specialised missile troops primarily armed with bows or similar long-range weapons may also be classed as close formation. Most of these are supporting troops found in the rear ranks of units primarily equipped for hand-to-hand fighting and who must conform to the formation of the majority. Some may instead be poor quality archer levies, clustering nervously together for moral support.

The amount of armour worn by close formation infantry varies widely, which makes it necessary to sub-divide them further. The present convention is to class those with an iron mail or bronze scale or plate corslet as HI (heavy infantry) and those with leather, quilted or lesser protection as MI (medium infantry). Similar metal protection to that of HI but extending to cover the lower arm and legs can upgrade to EHI (extra-heavy infantry). The ultimate grade is SHI (super-heavy infantry), represented in Ancients as opposed to Mediaevals only by dismounted Cataphract cavalry.

Most close formation infantry carry shields, the odd exceptions being missile troops or men needing both hands for a two-handed weapon, so not being able to use a shield at the same time.

Dispersed formation infantry are skirmishers. They are armed with javelins, bows, slings or other missile weapons. They shoot with these from a safe distance and try to evade charges by a fast sprint in the opposite direction, returning to shoot again when the chargers cease pursuit. Only very rarely will they try to fight hand-to-hand, and then only in difficult terrain. They wear no armour and often have no shields. This naturally helps their mobility, especially in woods or marshes or over hills, and they are often sent to seize commanding terrain in advance of the main body. They are also useful for supporting cavalry or countering elephants or chariots. They are all classed as LI (light infantry).

Loose formation infantry are a compromise between the others. They are mainly hand-to-hand fighters, but are generally lighter equipped and more mobile. Their typical weapon is javelin backed by sword, but thrusting spear or two-handed cutting weapon are also found. Some even carry bow or sling. They are much more mobile than close formation infantry, especially in difficult terrain, but cannot stand up to them very long in hand-to-hand fighting. However, they easily sweep away dispersed formation infantry, and stand a reasonable chance of catching them when they turn to run. Those that wear a light mail or scale corslet are classed as LHI (light-heavy infantry), those with lesser or no armour as LMI (light-medium infantry).

Mounted troops are divided into cavalry, camelry, chariotry, elephantry and mounted infantry, of which cavalry are the most important. Like infantry, cavalry can also conveniently be split into close, loose and dispersed formation types.

Close formation cavalry must restrict their speed to a trot if they are to retain cohesion. As they cannot rely on velocity to maximise the shock of impact, they must instead place reliance on weight. Such troops were called Cataphracts in Ancient times. We now class them as SHC (super-heavy cavalry). They wear metal armour covering their whole bodies including faces, hands and feet, and ride big horses armoured down to their knees. Their primary weapon is a long heavy lance called a kontos. This is too cumbersome for effective use after the initial impact, so they then take to sword or mace. They are so well armoured that they rarely bother with a shield.

Dispersed formation cavalry are primarily skirmishers from a distance. They are mostly armed with javelins or bow, and obviously have a better chance of evading a charge than LI. However, they are often willing to charge with the sword if heavier opponents become disordered or demoralised, or are otherwise at a temporary disadvantage. They have no respect for LI and will always sweep them away unless they take refuge where horses cannot follow. They are classed as LC (light cavalry).

Loose formation cavalry are the most common variety. Most of them are primarily armed with javelins, many with lance instead, and some add or substitute a bow. Those with distance weapons can skirmish in the same way as LC, but all are willing to charge home at the gallop to use lance, sword or light axe or mace. They are the most flexible troops of the Ancient world. Their armour varies widely. Those with non-metallic armour or depending entirely on their shields for protection are classed as MC (medium cavalry), those with mail or scale corslets as HC (heavy cavalry) and those with complete mail or with partially armoured horses as EHC (extra-heavy cavalry). Another class, SHK (super-heavy knight) is restricted to the mediaeval period. They wear complete armour of steel plates and mail and their horses usually have all-round protection to the knees, though often only of textile armour.

Camelry are something of a rarity. They are used by Arab raiders who cannot afford a horse, but even these usually dismount to fight as infantry. An occasional eastern general took advantage of horses' dislike for camels by using ridden or pack camels to disorder enemy cavalry, but this was very much a one-off gimmick. Most of the cavalry classes are duplicated for camelry by adding a small 'm' as a suffix, for example MCm (medium camelry). The big disadvantage of camels as opposed to horses is their clumsiness which makes them hard to manoeuvre and unsuited to skirmishing. Their advantages are cheapness to some armies, ability to go three days without water and cross soft sand, and strategic mobility. They are slower than a horse but can keep going longer, so may travel further in a day.

Chariotry was around before cavalry and was largely superseded by it, but was retained by some nations as specialised shock troops and by others apparently through lack of contact with other nations to bring home reality. We distinguish between lightweight chariots drawn by two horses and manned by an unarmed driver and single fighting man, which are analogous to LC and so are classed as LCh (light chariotry), and those with more fighting men or horses and mainly suitable for close combat, which are classed as HCh (heavy chariotry). One special variety of the latter has no fighting men but is fitted with scythe

Ancient warfare

A close up of the Dixon's Miniatures Turks, in the formation called 'Cantabrian Circle'. The figures are glued to card bases covered with dyed sawdust sold as a scenic accessory by model railway shops and fixed by dilute water-soluble PVA glue.

blades and intended for semi-suicidal charges to break up an enemy formation and create opportunities for other troops. This was the type to survive longest and was even used in one mediaeval battle between Byzantines and Normans. Deadly in ideal circumstances, it was relatively easily countered.

Elephantry originated in Indian armies, but spread westward after Alexander the Great's expedition to India. The Ptolemaic and Carthaginian armies lacked access to supplies of Indian elephants and substituted the smaller home-grown African forest species, the really big African bush elephant being too difficult to train. Elephants usually carried a crew of fighting men in addition to their driver, sometimes sitting astride, but later in a wooden tower or howdah. Regardless of breed or crew, we classify all as El (elephantry). Their main function is to charge into and disrupt an enemy formation, the elephant doing most of the fighting and its crew mainly seeking to protect it from enemy sneaking round behind. The most usual number of specialised fighting crewmen is two in addition to the driver if sitting astride, three or four if in a howdah. It has been suggested that the smaller African elephant may have had no crew apart from its driver. A secondary use exploits the dislike and fear which elephants cause horses not used to them. The elephant stays more or less out of harm's way but goes close enough to disorder enemy cavalry and thus help your other troops fight them. The disadvantage of elephants is that they don't like being hurt, panic easily, and don't care who they step on in the process.

Many a tabletop army has been left looking somewhat untidy after its own elephants have routed throught it.

That leaves artillery and ships, both probably better dealt with elsewhere.

Weapons

Because of the huge number of different weapons used during what is, after all, the longest of the wargaming eras, it has been necessary to gather them into groups with approximately the same uses and effects. Some obviously fit in more easily than others, but it is best to resist the temptation to increase the number of groups to cope with apparent exceptions rather than get bogged down in excessive complications.

Distant weapon classes comprise in order of descending range, stone or bolt-shooting mechanical artillery, crossbows, longbows, staff slings, ordinary bows, ordinary slings, darts, javelins, and chemical flame or bullet projectors. Some of these such as javelins, darts and bows are much less efficient against armoured than unarmoured targets. Others such as longbows and slings suffer less, and others such as artillery and crossbows practically ignore armour but are slow firing. Each class has its own advantages and disadvantages.

Cavalry weapons for close fighting come down to two classes, lance, and sword or equivalent. Infantry classes are two-handed pike, long thrusting spear, heavy throwing weapon, two-handed cutting weapon, and sword or equivalent. However, troops armed with most of these also get a bonus if they have javelins as well. Light spears that are not normally thrown are counted as javelins, but in close combat only. As with distant weapons, each class has its advantages and disadvantages. Pikes enable much deeper formations to be used, but suffer more than others if disordered. Two-handed cutting weapons are very good against heavily armoured enemy but prevent a shield being used at the same time. Long thrusting spears are better at stopping a cavalry charge than heavy throwing weapons, while the latter do well against infantry but can only be used in the initial contact, recourse than being had to swords. They are classed as close fighting weapons because they are only flung immediately before contact.

If one weapon system had historically been superior to all the others, it would have replaced them. This did not happen, so we must assume there was no super-weapon, nor must there be on the wargames table. The skill of general or wargamer lies in using his troops so as to maximise the advantages of their weapons over those of the enemy.

Morale and training

Having considered formations, armour and weapons, it is time to look at the men themselves. Our primary division is into regular and irregular troops. It is usually quite easy to decide whether a given force should be classified as one or the other, but quite another thing to define the two terms. One of the best ways of putting it is to say that an irregular warrior fights independently with his weapons with comrades around him doing much the same, while a regular officer fights with his unit, using his men as his weapons.

Regulars are more likely to wear uniforms and to be paid. Those fighting in company carry identical weapons and are likely to wear the same armour. They are commanded by officers instead of chieftains, and obey orders like 'Move to the left in column of threes' rather than 'You lot please go over there'.

Irregulars often have a mixture of weapons or dress, think of drill as ignoble

Ancient warfare 29

and demeaning but practice with their weapons, prefer to be persuaded rather than ordered and respect only their own chieftains. They are more likely than regulars to suffer a catastrophic drop in morale when things go badly, but they are also more likely to do something heroic beyond the normal call of duty. Except for light troops, they manoeuvre slowly and clumsily and fall easily into disorder. Their most effective tactic is often a fierce charge straight at the enemy. They should not be thought of as inferior to regulars, but as different. They often do better.

Regulars and irregulars alike are divided into four morale classes. 'A' Class regulars are spit-and-polish guardsmen of the most proud and self-confident sort. Their irregular equivalents are likely to be religious fanatics or beserkers.

'B' Class regulars are élite troops of higher than average efficiency. The irregular equivalent is a warrior of noble blood who in some circumstances may prefer death to dishonour.

'C' Class regulars are typical well-trained soldiers. The irregular equivalent can fight, but may be more interested in loot than abstract honour.

'D' Class regulars are second raters lacking full training and confidence and generally happier behind defences. The irregular equivalent is a more or less untrained peasant, usually conscripted at short notice and watched carefully—sometimes as they cross the horizon. However, properly positioned and protected they can do a useful job, and are dirt cheap.

Points values

Each model figure is allocated a value based on the amount of armour worn, the number of weapons carried, whether regular or irregular, morale class, and whether mounted or on foot. The cheapest is a 'D' Class irregular MI, LMI or LI with one weapon and no shield, who costs 1 point. The most expensive in any of my armies is an 'A' Class regular SHC with one primary weapon and a shield who costs 20 points. A similar system covers chariot, elephant and artillery models. Extra points are then allocated for command. Each regular unit costs 10 points, each irregular unit 25 points, a subordinate general 50 points and the senior general 100 points. The extra unit but lower individual cost for irregulars makes them cheaper overall than regulars if organised in big clumsy units, otherwise much the same.

National Convention armies are allotted a total of 1,000 points. 1,500 points is often prefered in other circumstances. 3,000 points has been used, but such a game may last all day and leave the players exhausted.

Choosing your army

Once you have decided which nation you are going to adopt and the point in its history that your army will represent, all is fairly plain sailing, for the Ancient player has the advantage compared with some periods of being able to buy booklets of army lists tailored to the rules and a series of books covering all the details of troop types, organisation, tactics and history that he needs to know. The lists are designed to allow considerable freedom of choice while still maintaining the essential character of the army. They were introduced in response to fervent requests by players and clubs who wished to curb individuals who relied somewhat too heavily on gimmicky troop choices to compensate for their lack of generalship. They are nevertheless not entirely accepted. If you come across a player whose army does not conform, he will claim that his own

The Romans are coming! An Ancient British warband musters to fight off the invader. The huts are made of Plasticine, the palisade home-cast from metal. The hills are carved from plastic foam, then covered with PVA soaked paper towelling and scenic flock or sawdust. The carpet tiles covering the table are especially good with 15 mm and smaller figures and can be used to give a patchwork field effect

research supersedes that of the published lists, and in some cases this will be true. All, however, agree they are a good guide for beginners.

Choosing nation and point of history is a very much more personal thing, though the lists and books can help by giving you some idea of the main possibilities. Basically, you should choose those that you feel a sympathy with, or as someone put it 'an army you can love even when it loses'. For lose it probably will until the beginner gains experience. Don't let this worry you too much. You will learn more from losing to a good player than from beating one even worse, though the latter occurence may well boost your morale very nicely. There is *always* someone worse, and sooner or later you will find him.

Some players discard this emotional choice in favour of picking the army currently showing the best record of wins in the Society of Ancients league championship or in their local club. Unfortunately, there are two fallacies here. The first is to believe that the rule and list writers have not been trying to make all armies approximately equal. The second is that the army currently doing well may be doing so because it is being controlled by an especially competent player who knows how to use it. This was recently brought out in a survey of Society of Ancients championship results, which showed that the top scoring army in the hands of the top graded players was almost the worst in the hands of beginners.

Although there is no army with overall superiority to all others, there is a certain amount of 'scissors cut paper, stone blunts scissors, paper wraps stone'. For example, an army long in cavalry and short in distant effect missile weapons has more to fear from an army strong in elephants than has an army

Ancient warfare

predominantly of infantry with plenty of missiles to annoy the jumbos at a distance. For this reason, some players have several armies and try to select the one best suited to deal with that of their opponents. Besides being expensive, such a policy has obvious limitations if the opponent also has several armies. A better policy is to stick to the same army and learn how to make effective use of it against any kind of opposition. Some of the armies doing best in the survey mentioned earlier did well in the hands of good players *because* they had acquired a bad reputation in the hands of others.

While there is no best army generally, there may well be a best army for you personally, one that you have an instinctive sympathy for and whose characteristics suit your particular talents. For example, if you think of the legions of Rome as irresistibly bringing civilisation to savage races, you will do better with an Early Imperial Roman army than with Ancient Britons. If you think of the Romans as unprovoked imperialist aggressors and the Celtic nations as artistic, freedom-loving individualists, the converse will apply. The following section will try to tempt your fancy by displaying thumbnail sketches of some of the hundred-plus Ancient armies currently available.

Some possible armies

Ancient British: Mainly relies on javelin-armed light medium infantry, best employed in an all-out enthusiastic charge, preferably downhill into disordered opponents. Supporting troops include light chariots to frighten and create disorder, light javelin cavalry to encircle flanks and exploit opportunities, light infantry with slings or javelins to distract or to 'shoot-in' an attack, and possibly even naked woad-painted fanatics to lead a desperate forlorn hope. Brightly coloured and fun to paint, each individual different, and with plenty of gay striped and check patterns. Especially suits a decisive-minded player prepared to chose his moment then fling the whole army in at once. Not for the ditherer who likes to hang back and skirmish.

Early Imperial Roman: Mainly relies on heavy infantry with heavy throwing spears, but backed by light-heavy infantry with javelins or bows and by javelin-heavy cavalry. Can also have lighter irregular barbarian allies, or bring light artillery into the field. Their armour, weapons and regular discipline make them the prime example of an offensive infantry army. A successful player needs determination rather than subtlety. Not especially worried by elephants, but can have difficulties with cavalry armies, the best tactic against these being to press forward and constrict them in a confined space against difficult terrain or a table edge. Not for a defensive-minded player.

Ancient Egyptian: A regular army whose main strength lies in its large quantity of disciplined chariotry. Has next to no cavalry, but plenty of supporting infantry, usually unarmoured and armed with light spears or bows. If attacking, its light chariots will need to be 'shot-in' by the infantry archers, who in turn need protecting by the spearmen. If acting defensively, its firepower will serve it in good stead against an inactive opponent, and the chariots can be kept out of harm's way ready to charge attackers' flanks. Its greatest strength is that most players are excessively nervous of chariots, so that a clever player can bring psychological pressure to bear by subtle manoeuvres. Its weakness is that its firepower encourages static linear deployments which are more vulnerable than a weak player thinks to either a well co-ordinated frontal attack or flanking. The infantry tend to look a little plain, although their appearance can be enhanced by incorporating various subject race contingents. The chariotry by contrast can look spectacular.

Ancient Greeks: The earliest Greek armies consist almost entirely of heavy infantry with long thrusting spears. Later armies incorporate good light and light-medium infantry and some rather more doubtful cavalry. However, the main strength still lies in the spears of the Hoplites, and a player who increases his supporting troops too much at their expense will suffer for it. Much like the Roman army in its use, but a little lower in offensive power against infantry, correspondingly better in defensive power against cavalry, especially if the flanks can be secured. A very pretty effect can be obtained in the mass by quite simple painting, but choice of figures is especially important.

Macedonians and Seleucids: The original Macedonian army of Philip and Alexander the Great has two main arms. The first is a large number of disciplined medium infantry armed with a pike that substantially outreached the Hoplite spear. The second is a smaller number of good charging heavy cavalry armed with lances but not carrying shields. These are supported by light cavalry also with lances, and by the usual range of Greek supporting troops. As a result of Alexander's campaigns, his successors changed the army by upgrading the pikemen to heavy infantry, introduced heavier cavalry, elephants, scythed chariots, horse archers and masses of poor quality Asiatic foot archers. The flexibility of choice and range of arms in such a Seleucid army has made it a favourite of many top players. However, its complexity makes it unsuitable for the less experienced, who do better with the simpler but still formidable original version. This army suits an offensively minded player with good timing. It is one of the few where simultaneous engagement of all units does not pay. The pikemen and light troops must fix the enemy's attention and create an opportunity for the heavy cavalry to strike. Because of its lack of shields, this cavalry is not at its best against enemy cavalry prepared to receive it, but is almost ideal for charging infantry, especially infantry already engaged with a steamroller pike phalanx.

Indians: The mass of an Indian army consists of infantry armed with powerful longbows and two-handed swords. However, its strength lies in its elephants and to a lesser extent, its heavy chariots. It has cavalry, but these are not of marked efficiency. Enemy cavalry armies suffer badly from elephant-induced disorder, and infantry armies are vulnerable to a charge of elephants and chariots, 'shot-in' by the infantry. The disadvantages are that the elephants are rather vulnerable to enemy archers, and that the army finds it difficult to change front to meet an outflanking movement. The army is naturally very colourful.

Sassanid Persians: The main strength of this army is its extra-heavy noble cavalry with bow and lance, equally able to shoot at a distance or charge. These are backed by a few elephants, super-heavy cavalry, light cavalry, light infantry archers and some extremely unenthusiastic levy spearmen. The combination of extra-heavy cavalry and elephants makes it extremely difficult for other armies relying primarily on cavalry to beat. However, being irregular, it will often beat itself by charging into dangerous situations without waiting for permission from its general! A good player can so manoeuvre it as to minimise the temptation. The bad player's most usual mistake is to include too many supporting troops at the expense of his main arms, and to let units be defeated in detail. When enough of an irregular army's bad units have run away, the rest often don't feel like charging. A very colourful army, each nobleman and his horse being in different bright colours and rich armour.

Parthians: A very simple army in composition, usually consisting entirely of super-heavy cavalry lancers and light cavalry horse archers. If used cleverly in combination, the archers can prepare the way for an especially devastating charge by the lancers, or the lancers protect the horse archers from molestation by enemy cavalry. If not so cleverly used, the horse archers may find themselves being ridden down while the lancers charge into thin air or ruin themselves against a pike phalanx. Not quite as colourful as Sassanids.

Huns: Even simpler, unless they have dragged some of their conquered subjects along. With the exception of a few nobles, the whole army will otherwise consist of light cavalry armed with bow and light spear. This is the most mobile army of all and will often run rings around the opposition. Unfortunately, they often won't do it much harm, although a moment's carelessness on the part of the other player can lead to almost instant disaster for him. Variegated rather than colourful, the painter trying to achieve an effect of menace and lack of washing facilities rather than prettiness. A fun army though.

Late Romans: Gibbon would have described these as decadent. They would probably have described themselves as up-to-date! Compared with the earlier army described above, they use much more cavalry including super-heavies and horse archers. The infantry have lost their metal armour in favour of moulded leather or none at all, but carry big oval shields painted in colourful regimental patterns, and have added a variety of light throwing darts of very long range to their earlier weapons. Each unit includes a proportion of archers. High quality regular troops can be combined with a variety of barbarian irregulars. A good all-round army, and my personal favourite.

I would like to continue with Assyrians swooping like a wolf on the fold, Germans swarming over the Rhine or Dacians down from the hills, painted Picts, inscrutable Chinese, fanatic Japanese, heavily armoured Tibetans, rampaging Vandals, sneaky Byzantines, charging Sarmatians, preaching Arabs and many others, but there is no room, so I shall have to refer you to the bibliography instead.

Special hints on recruiting, painting and basing Ancients

When buying your figures, bear in mind that, whilst regular close formation troops usually look best if all a unit's figures except officers, standard bearers and musicians are identical castings, this is not necessarily the case. For example, Greek Hoplites provided their own helmets and body armour, so you could use different figures provided the pose was the same. Lamming actually provide a Hoplite with interchangeable heads. When you come to troops fighting in loose or dispersed formation, you can enhance the appearance by bending arms a little so that shields and weapons are carried at slightly different angles. Light infantry do not usually shoot in volleys on command, so you can often mix figures from different manufacturers in slightly different poses. For example, one manufacturer may have an archer reaching for an arrow, while others depict identical men at full draw or having just loosed. Slingers should have right arm and sling bent to various angles. The same type of consideration also applies to later periods covered in this book, although a greater standard of uniformity should apply from the 18th century onwards.

If this technique vastly improves the realism of a regular unit, it is absolutely

Sue's Ancient British general and her bodyguard. Some people might think that the diminutive Airfix Queen Cartimandua looks a little out of place with husky Lamming warriors, but she wins battles, and Sue isn't going to change her!

essential for irregulars, who show even less uniformity. You should as a matter of course get your irregulars from as many manufacturers as possible. On top of that, you will find that each has figures nominally depicting other nationalities which, however, differ very little from those you are trying to portray. Fit these in as well. Some manufacturers cast figures with open hands and sell separate weapons and shields, so these too can be used to ring the changes. One of the most magnificent examples of this technique that I have seen was provided by George Gush's Viking army, which went a stage further by incorporating figures carrying barrels, drinking horns, captive maidens, other livestock and treasure chests as evidence of its prowess!

Even the best drilled regular unit has never managed to convince horses to march in step, though they may well halt in an identical position. Since horses are nearly always cast separately from their riders, a certain amount of swapping around between manufacturers is possible here too. This is lucky, because some manufacturers produce horses of rather strange shape or of excessive size for the Ancient period. Even if forced to use identical horse models though, it is easy to provide variations by bending neck and tail slightly, or by cutting loose a hoof from its base and altering the leg position.

The main difference between painting Ancients and figures of some of the later eras is the need to paint armour. Firstly, don't use silver. Keep it for sword and spear blades, the edges of axes and the points of arrows. Ideally, you should represent iron helmets, iron plate or scale armour, spear sockets and axe blades with a paint such as Humbrol HS 217 Steel. At a pinch, you can substitute silver with a touch of black added. Experiment a little first. Iron mail is darker still. Humbrol 53 Gun Metal is very good. Again, you can mix black with silver to produce a dark metallic grey. The various bronze paints sold are of doubtful utility as nearly all ancient bronze work was polished up to look like modern brass. So save your Antique Bronze for the occasional barbarian who does not look too fussy, and instead use Humbrol MC 18 Brass. Since some of the metallic paints don't dry too quickly, I personally apply them last, after varnishing. Some armour looks better if you apply a final thin wash of black to settle out in the depressions between scales or plates and pick them out. Scale armour especially benefits. Diluted ink will do.

The unskilled painter's maxim, 'If you can't be accurate, be complicated', applies with special force in Ancients. Paint an army of Picts with individual tartans for their cloaks and individual woad tatoos and no one is going to notice your mistakes! All irregular figures should be given their own individual colour scheme. This is not as time-consuming as it sounds. Having prepared a colour, you use it for the first figure's trousers, the second's tunic, the third's cloak, the fourth's hat, the fifth's bow case, and so on.

Even regulars need some painting variation, as units are not recruited from men of identical hair colour or ride identically coloured or marked horses. Decide on your norm, then mix a little black or sand coloured paint in progressively to give darker and lighter shades.

The size of the card bases on which you mount your figures must depend on the scale of the figures used, as this will change the ground scale. Under WRG rules each figure represents 20 men in four ranks of five, and each model of an elephant, chariot or engine a single rank of five. The frontages and depths allocated to men and animals are mainly based on surviving manuals, with interpolations where necessary. This enables us to duplicate real life formations quite accurately.

When mounting irregulars on their bases, avoid too much regularity. Another difference compared with other periods is that some cavalry are better mounted on individual bases instead of on multiple ones. This is so that they can use line ahead and circular formations.

Rules for Ancient wargaming

The playing rules presented in this book are neither designed to be comprehensive nor can they claim to be particularly original in concept. What they will do is give the beginner a taste of the flavour of each period without bogging him down in a welter of unnecessary detail whilst he or she is still learning. They are designed to produce fairly fast and furious games for the relatively small number of model figures that the beginner will initially have at his disposal. For those wishing to progress, having developed a liking for one or more particular periods, there are a large number of rule books available, along with suggestions for further reading in the bibliography. I must stress that all responsibility for these playing rules is my own and acknowledge the fact that several of the other contributors to this book do not entirely agree with them, so any comments, queries or criticisms should be addressed to yours truly! BQ.

So far as possible the rules are designed on a modular principle, in that at the end of each chapter there is a section which builds on what has gone before, rather than repeating the basics each time. This will lead to some anomalies when you come to progress to more advanced sets of rules as the man to figure ratios, ground scales, unit frontages, march rates, etc, will in many cases be different. For this reason I would strongly recommend that at this stage you do not permanently affix your troops to cardboard movement bases but simply Blu-tack them in place.

The basic premise upon which the following rules are based is quite simple: that, although cultures and methods of waging warfare have changed drastically over the centuries, the essential physical capabilities of an infantryman or a cavalry horse have not. As any anthropologist will realise, this is not strictly true. Man has evolved, even over the last 2,000 years. He has gradually grown taller and the realisation of the need for hygiene and dietary supplements have increased his average lifespan. Moreover, in a military context, different armies within even the same historical period have frequently adopted different drill and march rates. However, for our purposes a medium infantryman (MI) in 1800 BC is assumed to be carrying about the same weight of weapons and kit as his counterpart in 1800 AD. It is only when we come to consider the complexities of mechanised armoured warfare in the final chapter that a radically new system becomes necessary. However, on to the rules.

1 All movement is simultaneous, as explained in the Introduction.

2 Each model infantryman or cavalry figure represents 20 actual men. Each model catapult, ballista or, later, cannon, represents two actual pieces and is crewed by two figures; when one figure is eliminated the model only represents a single piece and its ability to inflict casualties is therefore halved.

3 Each game move represents one minute of actual combat.

4 One inch on the tabletop represents ten paces (roughly 25 feet) and all movement rates and weapon ranges are measured in paces.

Sequence of play

1 Write orders.

2 Declare any charges. (Cavalry which have been given the precautionary order 'countercharge if charged' may do so, otherwise they must receive their opponent while remaining stationary.) Test morale of troops charging or receiving a charge. NB, skirmishers may attempt to evade a charge rather than receive it.

3 Preliminary missile fire. (Since the game move represents a full minute of actual battle time, firing is actually continuing throughout the move, but for the sake of game convenience is divided into two phases. The preliminary phase will normally only affect the longer-ranged weapons—from bows and crossbows up; throwing javelins and the like will be utilised in the second firing phase.

4 Movement. (Move any units you wish up to their total movement allowance, apart from those firing, who must remain stationary.)

5 Second firing phase. (Throw javelins, axes and other short range weapons.)

6 Mêlée. (Fight any hand-to-hand encounters.)

7 Morale. (Test the morale of any units necessary according to the ensuing table.)

Return to phase 1.

In parentheses, it should be understood that, since all movement and firing is simultaneous, no casualties are removed until the end of phase 5, followed by any losses sustained in the mêlée phase 6.

Movement

As noted above, all movement is in paces, there being ten paces to an inch. Thus a troop type with a normal movement factor of 120 may move up to 12 inches per move. The abbreviations following are explained by Phil in the preceding chapter.

Troop type	Normal	Charge
SHC	80	120
EHC, HC & MC	120	160
LC	160	200
HI & MI	60	80
LI	80	100
HCh	80	160
LCH	120	160
Elephants	80	120
Cm	120	120
Light artillery	60	—
Heavy artillery	40	—

Charge moves may only be made if they have been declared (phase 2 above). Their advantage is that they give the charger added impetus in the ensuing mêlée. Troops ordered to countercharge if charged must delay for a quarter of a move's reaction time and can thus only move up to three-quarters of their normal charge move.

Effects of terrain on movement

All normal move rates are halved and charging is impossible over steep hills, amongst buildings, in woods, marshes or sand or over streams, low hedges and walls. Artillery can only cross walls through a gate or breach. Only infantry can operate within buildings. Troops on paved roads running through difficult terrain move at their normal rate except up steep hills.

Interpenetration

Troops in close formation, whether cavalry or infantry, cannot interpenetrate each other without both units becoming disordered. Troops in open formation may interpenetrate at will without effect. This means, for example, that archers—or, later, musketeers—wishing to evade cavalry or a similar threat behind the safety of a pike block, may only do so if the pikemen open their files to extended order. Of course, they could alternatively shelter *under* the pikes without disordering the formation.

Formation changes*

Close order to extended order and vice versa—$\frac{1}{4}$ move.
Increasing or decreasing unit frontage by two figures—$\frac{1}{4}$ move.
All other changes—$\frac{1}{2}$ move.
Altering front more than 30° and less than 150°—$\frac{1}{2}$ move (0–30° and 150–180° changes are virtually instantaneous).

There are no penalties for altering a unit's stance; ie, pikemen can go from grounded pikes to the 'kneeling' stance required to receive cavalry, or archers load and fire, virtually instantaneously.

Disorder

Troops become disordered or 'unformed' if interpenetrated whilst in close order; if engaged in mêlée whilst changing formation; if engaged in mêlée from

---

*These factors apply only to regular troops; poorly trained militia or irregulars take double the time.

flank or rear; after a round of mêlée in any case; whilst routing or pursuing; crossing walls, hedges or streams; on steep hills except on a paved road; all except light infantry inside buildings; all except light infantry in woods or marshes; formed cavalry charging down steep hills; *in other circumstances according to common sense or the umpire's discretion.*

The last is an important point to note. No simple set of rules can cover all conceivable battlefield circumstances, so it is up to the players to reach common agreement when a situation arises which is not covered. Hopefully, this can be done amicably. If an umpire is present, his decision must always be taken as final, although it is, of course, quite permissible to argue with him *after* the game.

Base sizes

The following are the figure frontages recommended by WRG for 25 mm figures and should be halved for 15 mm. SHC, HI and MI—15 mm; EHC, HC, MC and Cm—20 mm; LC and LI—30 mm; elephants, light artillery, LCh—40 mm; HCh—60 mm (four horses); heavy artillery, siege towers, etc—80 mm. It is important that figures are correctly based so that they obtain the appropriate advantages and disadvantages in terms of movement, formation changes, vulnerability to missile fire and effectiveness in mêlée.

Regardless of type, no unit should be smaller than four or five figures (80-100 men) or larger than 50-60 (1,000-1,200 men), and should where possible be based upon historical precedent. Obviously, a book of this nature cannot possibly cover all of these, but a good average to aim for is between 200 and 1,000 infantry (10 to 50 figures) and 100-300 cavalry (5-15 figures). In the Ancient period the proportions of infantry to cavalry and missile weapons to shock weapons varied widely, and WRG publish a booklet of army lists based on a points system which gives all the most popular alternatives. In fact, for anyone wishing to take 'Ancients' seriously, the WRG playing rules, army lists and reference books listed in the bibliography are invaluable.

Missile fire

As noted above, this takes place in two phases for convenience: firstly the long-range fire, and secondly the close-range. For this purpose, close range is taken as being anything under 50 paces, long range anything above. Ranges are very simply memorised as follows: close-range weapons (ie, javelins, throwing axes, etc)—40 paces; slings and horse or otherwise mounted archers—120 paces; staff slings, bows and crossbows—240 paces; artillery—480 paces. These figures ignore some finer distinctions which you will encounter when you progress to other rules. Units may fire up to 30° to either side.

Since all missile fire until the advent of mechanised and electronic fire control systems was essentially line-of-sight, no firing overhead is allowed except that archers may fire over one rank of the same unit of cavalry or chariots, or two ranks of infantry, while elephant crews and all troops on a higher level of a hill may fire (and be fired upon) over the heads of any intervening troops.

Other than when throwing close range weapons, units firing must remain stationary throughout the game move in order to do so. In other words, you may move *or* fire but not both. The exceptions to this are elephant or chariot archers, camelry and cavalry. When firing these troop types, the range is 120 paces; in other words, elephant, camel and chariot mounted archers count as cavalry insofar as the firing overhead rule is concerned.

It is permissible to fire into a mêlée between friendly and enemy troops; however, casualties must be divided between the two in proportion to the respective numbers engaged on each side. Friendly troops treated in such cavalier fashion will, needless to say, suffer under the ensuing morale rules.

Each weapon type has a different factor depending upon the type of target so that, for instance, a light javelin is less effective against a heavily armoured man than against an opponent with no armour. The following figures are based, although slightly amended as a result of personal preference, upon those included in the WRG rules. Abbreviations as hitherto.

Weapon v:	SHC	EHC	HC	MC/Cm	LC	HI	MI	LI/Art	HCh	LCh	El
Hand-hurled weapons, slings and bows*	0	1	2	3	4	1	2	3	2	1	1
Crossbows and staff slings	2	3	3	3	1	2	2	1	2	0	1
Artillery	3	3	3	3	1	3	3	1	2	1	2

In deference to WRG supporters, I must stress that the variations in the above table from the published rules are purely the product of personal preference resulting mainly from a desire to give more 'expensive' troop types in terms of points value a greater advantage. Historically, they may not be accurate, but they lead to greater thought being given to the disposition of your various missile arms to counteract those weapons available to your opponent.

Deduct 1 from all the above factors if firing overhead or if target unit is charging or has the advantage of cover (woods, walls, buildings, earthworks, etc).

Also add 1 if target unit is not carrying shields within 90 degrees of arc of fire (see diagram).

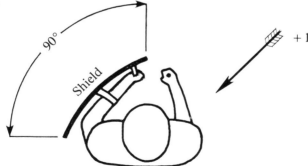

Note that troops utilising shields may not operate double-handed weapons, and vice versa, within the same move. In other words, a double-handed axeman, or an archer, may use his shield defensively *or* his weapon offensively, but not both simultaneously.

Casualties from missile fire

Take the above factors according to the weapon and type of target; throw an ordinary dice (1–6) if fighting with irregular troops or an average dice (2–5) if

*Excluding the Welsh/English longbow—see next chapter.

fighting with regulars. Add or deduct as above. The following table gives the number of enemy *men* (not figures) killed.

Score	Number of figures firing											
	1	2	3	4	5	6	7	8	9	10	11	12
-1	0	1	2	2	3	3	4	4	5	5	6	6
0	1	1	2	3	3	4	4	5	6	6	7	7
1	1	2	2	3	4	5	6	6	7	8	9	10
2	1	3	4	5	6	8	9	10	11	13	14	16
3	2	3	5	6	8	9	11	13	14	16	18	19
4	2	4	6	8	10	12	14	16	18	20	22	24
5	3	5	8	10	13	15	18	20	23	25	28	30
6	3	6	9	12	15	18	21	24	27	30	33	36
7	4	8	12	16	20	24	28	32	36	40	44	48
8	5	10	15	20	25	30	35	40	45	50	55	60
9	6	12	18	24	30	36	42	48	54	60	66	72
10	7	14	21	28	35	42	49	56	63	70	77	84

In order to keep a record of casualties, make a note of each unit's name or number on a pad of paper and, each time it comes under fire, write down the casualties inflicted. You only remove a model figure when 20 men, or a multiple of 20, have been killed, and then only after phase 5 or 6 of the move as noted above. When using this table, an artillery piece counts as two 'figures' until reduced to half strength.

Mêlées

The above table is also used in calculating casualties from mêlées, for the sake of simplicity although in fact the proportions *should* often be different, as you will find when you progress to some more sophisticated playing rules. As with missile fire, you first look at the weapon carried by your own unit, then the type of target, in the following table. You then throw a 1-6 dice for irregulars or a 2-5 for regulars, and *add* +1 if attacking from flank or rear, or if enemy is unformed, or if you are charging and he is not, or if he is falling back and you are pursuing, or if you are uphill; and *deduct* -1 if you are unformed or being pursued or the enemy unit has shields. (To give an example, if two formed bodies move into a mêlée, at the end of the first move they are both unformed—see above—so count +1 for fighting against unformed troops and -1 for being unformed themselves, or, in other words, continue as normal until one side breaks.)

Weapon v:	SHC	EHC	HC	MC Cm	LC	HI	MI	LI Art	HCh	LCh	El
Cav or Cm											
Long spear	2	2	2	4	3	3	4	5	1	2	1
Other	0	1	1	2	2	2	3	5	0	0	0
Inf											
Pike	3	4	5	5	2	2	3	2	3	4	1
Shorter spear	2	2	3	4	3	2	4	3	2	3	0
Two-handed weapon	4	5	5	5	4	5	5	3	1	1	1
Other	1	1	0	1	1	1	2	2	0	0	1

Add +1 to above factors if fighting from chariot or elephant.

Morale

Any set of wargames rules lives or fails according to the standard of its morale section. Battles are usually won as a result of people giving up rather than being killed, and historically it is rare for the casualties in any battle up to the First World War to exceed 25 per cent or so, although there are, of course, instances of individual units within a battle being totally wiped out. The object of morale rules is to tell the commanding player when his model soldiers have had enough as, lacking vocal cords, they cannot do so themselves!

You test the morale—or fighting spirit, if you like—of units under your command according to the following circumstances: when commencing or receiving a charge; after each move of mêlée; if it has received ten per cent casualties during one move or when it has lost 25 per cent of its original strength in any case (*and* every move thereafter); if the commanding general is killed (test for every unit within 300 paces); under other circumstances dictated by common sense or the umpire.

Regular troops throw an average dice, irregulars an ordinary 1–6 dice. To the score, add or deduct as follows:

Own troops advancing	+1	Own troops uphill from enemy	+1
Own troops standing	0	Own troops downhill from enemy	−1
Own troops withdrawing	−1	Own troops in cover	+1
Own troops routing	−3	For every ten per cent casualties from original strength of unit	−2
For each friendly formed unit within 150 paces	+1	For each secure/insecure flank	±1
For each friendly routing unit within 150 paces	−1	Each enemy unit to flank or rear	−1
For each enemy formed unit within 150 paces	−1	Enemy El or Ch within 300 paces	−1
For each enemy routing unit within 150 paces	+1	Own unit in 'uncontrolled charge'	+3
		General with unit	±2
Own troops unformed	−1	Fired upon by friends	−2
		General killed	−2

NB Units must be visible to affect morale.

The above can give a result from about −10 to +15, depending upon circumstances. If the result is a minus score, the unit throws down its weapons and routs, ie, runs away from the enemy at charge speed. If it is between 0 and 3 the unit will stay on the table but may not advance until/unless circumstances alter to bring its morale factor above 3. If it is between 4 and 8 the unit may advance or otherwise carry out its orders as normal except that it may not charge. If it is between 8 and 12 the unit will do exactly as ordered or desired. If it is above 12, however, the unit is assumed to have really got blood-lust, and will *charge* straight towards the nearest enemy unit, regardless of instructions or how stupid this may seem to the commanding player! This condition applies until the end of the first move of mêlée when morale must be tested again, and is called an 'uncontrolled charge'.

Note that all the above factors are accumulative; thus, a unit which throws 3 on the dice while advancing (+1) with two friendly units (+2) on either flank (+2) against an uphill enemy (−1) will end up with a total of 7 morale points and will thus continue its advance. The importance of keeping units well supported is thus clearly underlined.

Late Romans in action against their most dangerous traditional enemies, the Sassanid Persians. Persian super heavy and extra heavy cavalry advancing supported by elephants have been countercharged by Roman cavalry. The Persian infantry huddle miserably to the rear, while their light cavalry attempt an outflanking move. The Sassanid figures are obsolete Miniature Figurines.

General

There are many other factors which could be taken into account in a set of wargames rules, and the many commercially available sets *do*. However, the above is all that you really need for an enjoyable game and, particularly if you have the services of an umpire, can be adapted to take into account an endless variety of battlefield situations.

Chapter 2

Mediaeval warfare
by Ian Heath

The first problem one encounters when dealing with the mediaeval era is to establish satisfactorily when it began and likewise when it ended. To most people the word 'mediaeval' conjures up a picture of mounted knights encased from head to toe in a shell of metal plates, the armour itself being of a type which did not effectively become widespread until the 15th century. However, the thousand years of military evolution which separate the Roman legionary of late antiquity from the armoured behemoth of the late Middle Ages remain, to many, a total void in which just a handful of names or events sparkle through the darkness, such as the Battle of Hastings, Richard the Lionheart, Henry V and the Hundred Years' War. But what of the rest? It is undoubtedly this very vagueness that has relegated the mediaeval era to its status of 'poor relation' in wargaming terms, where it is branded as part of the 'Ancients' scene and is equally incomprehensibly largely covered by the same rules as those governing Ancient Greek and Roman warfare.

When did the mediaeval era actually begin, then? Opinions inevitably differ on this and range from 476 AD, when the Western Roman Empire fell, through 500, 600 and 800 (for no more logical reason than that they are nice round figures) right up to 1066—which is only applicable for the British Isles anyway—and 1071, when the Byzantines of the Eastern Roman Empire suffered their disastrous defeat at Manzikert in modern Turkey. The problem with all these dates, of course, is that they presuppose a nice clean break between the Ancient and Mediaeval worlds, which obviously did not exist. Realising this, more and more historians are now beginning to recognise a sort of transition period which, because of the total void, mentioned earlier, out of which is was originally born, goes under the name of the Dark Ages; its own particular parameters are circa 500 to circa 1000 AD, though these too vary. Like all transition periods it was a time of slow but revolutionary change, and right in its very middle, in the mid-8th century, lie the roots of the Middle Ages.

Feudalism

The very keystone upon which mediaeval military organisation was built was feudalism. The basis of feudalism was the granting of an estate called a benefice or fief in exchange for mounted military service by the recipient, and often a predetermined retinue of men in addition, for a specified period which varied according to time and place but was most commonly 40 days by the 12th century. Such grants were inevitably only made to those financially capable of adequately equipping themselves and their followers and performing the requisite service (which was at their own expense), so it comes as no surprise to find that in most European languages the words for a cavalry soldier and a noble gentleman became synonomous, such as English *knight,* French *chevalier* and Spanish *caballero* to mention but three examples. In strict feudal terminology, however, the holder of a fief was called a vassal.

In broad terms, then, feudalism was a system by which unpaid military service was made available in return for land. It seems to have first appeared in this form between 730 and 750 AD in France, when military service started to replace the agricultural or personal services which vassals had previously owed. Under the Emperor Charlemagne it became more widespread and various capitularies of the early 9th century give us details of how many *mansi* of land (basically manors, or estates) owed a man to the army. It was under Charlemagne's successors, however, that feudalism really flowered when constant civil wars had so weakened the central authority of the crown that the king was obliged to rely heavily on his vassals for military support. It had become the norm over most of Europe by the end of the 11th century and remained so for something like the next 200 years, though in some places, notably Scandinavia, Germany, Russia and Eastern Europe, feudalism never established itself to quite such the same extent as it did, say, in France and England. As an indication of the system's effectiveness, feudal tenants in 12th century England and 13th century Castile and Sicily each owed the crown the service of some 5–6,000 knights. France, on the other hand, although the birthplace of feudalism, was owed just about the smallest feudal service of all, because its nobility paid little more than lip-service to the monarchy and the great dukes and counts provided contingents of only a handful of knights each, even though their own vassals owed them the service of *thousands.*

Those vassals whose conditions of enfeoffment required that they bring with them to the army additional similarly armed men could either supply them all personally (which involved maintaining a standing retinue of household knights or hiring mercenaries), or could subinfeudate parts of their fiefs to other men likewise in exchange for military service, these lesser tenants then providing the tenant-in-chief's retinue. More often than not, in fact, tenants-in-chief were owed the service of greater numbers of knights than they themselves owed to the king, the extras being useful in baronial squabbles or else to ensure the tenant-in-chief's own safety in battle if he should deem his requisite quota insufficient for this purpose. Those who stayed at home sometimes contributed to the upkeep or equipment of those making up the stipulated feudal contingent. On the other hand, those who failed to turn up when required were heavily fined, it being cheaper in fact to announce beforehand that you were not going to appear in which case all you had to pay was *scutage,* a money-payment in lieu of service that was used to hire a mercenary substitute. One source explains that it was called scutage because it was paid according to the number of *scuta* or shields owed, the shield in this instance representing a man.

'Knights in armour'

It was the charge of the armoured knight that was the principal characteristic of mediaeval warfare, though admittedly this is more true of the 12th–14th centuries than of the rest of this era. In Carolingian times the cavalry had fought in a relatively loose, uncoordinated mass, hurling their lances javelin-fashion

An Agincourt-style reconstruction using Bill Lamming's superbly painted figures. English archers and men-at-arms are charged by French knights glorious in their heraldic splendour.

from a distance before charging in in small groups to exploit with their swords any momentary advantage that might present itself. The Bayeux Tapestry and contemporary accounts of the Battle of Hastings indicate that this practice was still current even in the mid-11th century, but by this time the lance was also being couched (the couch, with the lance held tight against the side under the arm, being a means by which the full weight of both horse and rider could be put behind the thrust). By the 12th century the knights advanced at a slow trot in tight formation, gathering momentum to a full gallop only over the last 75 yards so that the chances of losing formation or spreading out were minimised; what therefore hit the enemy—with a resounding crash, so the sources repeatedly tell us—was a single dense, armoured mass that, according to one Byzantine writer, was capable of punching a hole right through the walls of Babylon. This was the very essence of what has been graphically, and very accurately, described as 'mounted shock combat'.

Tactics

Battlefield tactics were basically elementary, although a good commander would pay close and careful attention to the nature of the terrain and the initial disposition of his forces. The knights generally formed up in close order in a number of divisions that are often described in the sources as *batailles* or 'battles', each usually comprised of several feudal contingents and/or mercenary companies. The formal arrangement was three 'battles', which on the march formed the vanguard, main body and rearguard. On the battlefield they could be drawn up in a variety of arrays, principally in echelon (with one flank—usually the left—advanced and the other refused), in line, or in column, although understandably there were also many subtle variations of these straightforward formations. Reserves were usually held back, sometimes concealed off to one side so that they could charge in on the enemy's exposed flank. Once battle was joined the individual feudal divisions tended to do their own thing; it was, after all, extremely difficult to rally or recall mediaeval troops once engaged, and on the whole a general could only anticipate launching his decisive cavalry charge once. It would either roll the enemy up on impact or else, by weight of numbers or dint of hard fighting, rout him in the ensuing mêlée. If the first attack failed but successfully rallied it was a sure bet that each additional charge would nevertheless be weaker than the last in both numbers and vigour.

It should not be assumed, however, that the knight reigned supreme on the mediaeval battlefield. For a start there were lesser feudal tenants called sergeants, two of whom were usually equated to one knight, and they similarly fought mounted but on less expensive horses and in less complete armour. More to the point, however, the knight, like his modern counterpart the tank, was expensive, difficult to maintain at full efficiency over prolonged periods, and required complex logistical arrangements and the assistance of various specialist troops in battle. Without the support of infantry, for instance, he was incapable of assaulting castle walls, and on the whole he could not hope to beat horse-archers like the Turks and Mongols or pike-armed infantry like the men of Flanders and the Swiss cantons without support from his own bow- or crossbow-armed foot soldiers. There were some specialist cavalry capable of providing firepower, such as mounted crossbowmen, the Spanish javelin-armed *ginetes* and the Turcopole horse-archers of the crusader states, but on the whole (the main exception being the Asiatic nomad hordes) it was essential that every army included at least some infantry to perform those military functions which were beyond the knight's capabilities.

Inevitably the infantry were, generally speaking, less well-protected than the knights, but the statements made by many modern-day authorities about 'poorly armed infantry with no internal or tactical organisation' are manifestly absurd. One book I read recently described mediaeval infantry as 'scorned by the knights and regarded as a kind of picturesque adjunct to battle, being on the field so that the knights might show their superiority by cutting them down in their droves'—and a more apocryphal view than that you could not hope to find anywhere! The truth of it is that although there *were* sometimes foot soldiers present who were no more than hastily mustered ill-armed rustics, mediaeval infantry were more often than not well-equipped militiamen, mercenaries or feudal retainers quite capable of giving a good account of themselves against cavalry. This was particularly true of infantry armed with double-handed weapons, long spears, bows or crossbows. The axe-armed Saxons at Hastings, for example, successfully held off repeated Norman cavalry attacks and were only defeated when William eventually brought up his archers against them in conjunction with yet another cavalry charge. Similarly at Falkirk in 1298 the Scots' *schiltrons,* bristling hedgehog-like circles of spearmen, bloodily repulsed the charge of the English knights but were then broken in turn by King Edward's English and Welsh archers. The lesson to be learned on both occasions was that if they lacked cavalry support of their own, irrespective of how well they were personally armed, the scales were weighted against an all-infantry army in battle with a combined force of horsemen and foot soldiers. That is not to say that an all-infantry army could *not* win, even on an open plain ideally suited to the enemy's cavalry, as was proved by the Swiss victory of Lauppen in 1339 where halberdiers and pikemen severely discomfited a large force of Burgundian knights and infantrymen.

The Crusades

The main theatre of war throughout the 12th and 13th centuries was in the Levant, where crusaders of umpteen nationalities carved out and held several principalities which were collectively known as *Outrémer,* the lands across the sea. It was here, more than in Europe, that the full potential of the foot soldier was first fully recognised. He formed a screen in front of the knights and mounted sergeants to protect their horses from the hail of Moslem arrows; he protected their marching columns like a wall, marching backwards in the rear ranks to fend off repeated Turkish attacks; and he discouraged the close approach of the enemy's cavalry by his accurate and steady crossbow-fire. At one battle in 1192 during the Third Crusade King Richard of England combined his infantry into a solid wall, the spearmen setting their shields in the ground and holding their lances out-thrust at the chest-level of charging horses, while between each two spearmen, behind the shelter of the shields, crouched two crossbowmen of whom one loaded while the other fired. So withering was the fire they laid down that the Moslem cavalry, despite overwhelming numerical superiority, refused to close with them, eventually taking flight in the face of a general advance by the infantry. They left 700 men and 1,500 horses dead on the field of battle, the most amazing thing of all (other than the English claim that they themselves lost only two men) being that, with the exception of—at the most—15 moth-eaten horses, the Christians on this occasion had no cavalry at all.

Although they managed to hold on to the Syrian seaboard in the face of overwhelming odds for the best part of two centuries, the crusaders failed in their ultimate aim of permanently liberating the Holy Land from the Moslems. After their decisive defeat at the Horns of Hattin in 1187, and despite the workmanlike efforts of King Richard's Third Crusade a few years later, it was inevitable that the fall of Outremer was only a matter of time. Despite the failure of the Crusades, however, they had nevertheless been of indirect value in a variety of other ways. Not least of all, they had attracted away from Europe many of its hawks, the adventurers and militants to whom fighting was second nature. More importantly, however, they had led to developments in strategy and tactics and given useful experience in the mounting of long-distance expeditions (both of which were lessons from which Europe failed to profit), as well as nurturing the skill and the cult of knighthood.

Arms and armour

During the same period there were also advances in military architecture, and in the technology of the armourer. Pieces of plate armour gradually began to supplement the mail hauberk which, apart from a quilted jerkin, had been the knight's sole body protection since the beginning of the mediaeval era. Now iron knee-cops and greaves replaced the old mail *chausses* or leggings, the all-enclosing barrel-helm with its perforated face-mask replaced the Norman-style nasal helmet, and an almost triangular shield (called a 'heater' shield) replaced the kite-shaped one which we know so well from the Bayeux Tapestry. Horses too started to wear armour again at about this time (they had worn it in classical times and Byzantine horses possibly still did), this comprising a body-piece called a *bard* and a head-piece called a *chanfron*. Over it all both man and horse wore loose linen coverings, called a surcoat and a housing respectively, on which by the middle to late 13th century, and until surcoats and their successors, tabards, went out of fashion in the 15th century, were displayed the individual knight's own colourful coat-of-arms, for by this time heraldry was in full bloom.

Although pikemen too were making a dent in the knight's battlefield supremacy from the 12th century onwards, particularly those from the Low Countries (notably Brabant and Flanders), it was the crossbow that was the infantryman's weapon *par excellence* during the Crusade era, both in the East and in Europe. Although the skill required in its production, maintenance and use restricted its availability largely to wealthy town militias and mercenaries, and despite a vehement Papal Interdict issued against it in 1139, and again on a variety of occasions thereafter, the crossbow flourished. Before the 13th century had progressed very far crossbowmen greatly outnumbered archers throughout most of mainland Europe, foremost among its exponents being the Genoese; crossbowmen in the French army at Crécy were Genoese, for example. England, however, was an exception to the rule, for we had by this time discovered a weapon that the French were later to laugh at—to their cost—as a 'crooked stick'. This was the longbow.

Contrary to what most people think, the longbow was not an English

Bill Lamming's English and French knights on the same side in this late Crusade period engagement against Phil Barker's Saracens.

Raid on a manor house in the Crécy period. 25 mm figures by Bill Lamming, the house being constructed by means of the Linka building system available from Thomas Salter Ltd.

invention but was in fact Welsh. Henry I had been the first English king to take a special interest in the bows of his Welsh mercenaries, but it was not until the late 13th century that this weapon really began to receive serious attention. The ground had already been cleared ready for its rise to fame by an Assize of Arms issued in 1242 by which most categories of English militiaman were expected to perform their military service armed with a bow, though this was still the earlier and less powerful short-bow. It was Edward I, who reigned 1272–1307, who can justifiably be credited as the father of the military longbow. He recognised its power and potential and, as well as hiring large numbers of Welsh archers (as many as 10,900 in 1298), he also set about training Englishmen in the use of this simple yet sophisticated weapon.

The longbow was made predominantly of yew, otherwise of elm or ash, and was some six feet in length—hence its name. Unlike the short-bow, which had been drawn to the chest, the longbow was drawn right back to the ear. It had a maximum effective range of some 350 yards, could apparently penetrate armour at some 250 yards, and a trained archer could shoot up to 15 reasonably aimed arrows a minute compared to the single bolt that a crossbowman could fire in the same space of time. Before long large contingents of English as well as Welsh archers were to be found accompanying every English army in the field, and they were soon accompanied in action by another innovation—dismounted knights.

Admittedly, dismounted noble horsemen had been present in many battles since the very beginning of the mediaeval era, either dismounting in desperation (since it is impossible for cavalry to stand on the defensive), or to bolster the less reliable and less well-equipped infantry. But in the 13th century this occasional usage had apparently died out. Now, under the auspices of Edward III, it was back with a vengeance, and the hideously efficient combination of archer and dismounted man-at-arms scored success after success at Dupplin Muir in 1332, Halidon Hill in 1333, Crécy in 1346, Poitiers in 1356, and even in a sea-battle at Sluys in 1340. From the 1330s right through to the 1420s this tactical combination was to prove virtually unbeatable, culminating in the most spectacular English victory of the Middle Ages when, at Agincourt in 1415, Henry V with 5,000 archers and 1,000 men-at-arms defeated an alleged 46,000 Frenchmen of whom, for the loss of no more than a few hundred men, they slew some 10,000 including the Constable, three dukes, seven counts, 120 barons and 1,500 knights.

The large-scale employment of the longbow had inevitably involved the introduction of a brand new battlefield formation which utilised its firepower to the best advantage. This was the *herce* or harrow, a hollow wedge of archers which projected in front of the battle line. The archers were usually arranged in several such *herces* between the bases of which were drawn up the dismounted men-at-arms, and additional archers were positioned on the flanks, which were angled forward, enabling the archers to enfilade the enemy as he approached. The French against whom this tactic was chiefly employed, it should be noted, failed to respond by any evolution of their own tactics. As one modern author has observed, 'the undisciplined French nobility, their heads filled with obsolete notions of chivalry, regarded all infantry opposed to them as a species of insult'. At Crécy their knights charged doggedly into the face of the English arrow-storm no less than *fifteen* times and suffered accordingly; at Poitiers, thinking that the lesson to be learned from their defeat at Crécy was to dismount their men-at-arms like the English had, they advanced in a ponderous wedge on foot when they would have done better to have remained on horseback; and at Agincourt, having 'forgotten nothing and remembered nothing' as one source succinctly put it, they once again charged hopelessly against the hailstorm of English arrows, their first two lines dismounted and their third on horseback. It was not an improvement in tactics that was to finally beat the longbow, but an advance in technology.

Men-at-arms

The observant among you may have noticed that in the last couple of paragraphs the term 'men-at-arms' has suddenly appeared. This is for a very good reason, which is that from the late 13th century onwards very few armoured horsemen, be they mounted or dismounted, were actually knights (ie, men serving in exchange for fiefs). The cost of being a knight, both in equipment and in social status, the latter of which involved chivalric displays of 'largesse' and a generally extravagant life-style, was by now becoming prohibitively expensive. As a result the majority of cavalry were now esquires either awaiting knighthood or not wealthy enough to accept it, retainers employed on a temporary basis in exchange for *fiefs-rente* (fiefs paid in cash rather than land), mercenaries, landless men of knightly families, or feudal tenants of less than knightly status. The fact that fewer actual knights were available led to the adoption of the value of possessions rather than the holding of land as the qualification for obligatory mounted military service. This ensured that those who served as 'knights' were best able to afford the ever more sophisticated armour that, in response to the threat of the longbow, grew steadily heavier and more complicated throughout the course of the 14th and 15th centuries, posing successive generations of armourers with the perpetual

A skirmish between Saxon fyrdmen and Norman knights. Hinton Hunt figures.

dilemma of those conflicting requirements, unrestricted manoeuvreability and increased protection. The term 'man-at-arms', applied indiscriminately to any warrior who fought in armour on horseback, blurred any remaining distinctions between those of knightly and those of non-knightly rank.

The subsequent alarming decrease in the number of feudal troops available was largely offset by paid retinues and mercenaries, both cavalry and infantry. Indeed, the constant warfare of the so-called Hundred Years' War, which lasted intermittently from 1337 until 1453, was responsible for a great increase in the number of mercenaries available, these being the *routiers*—'rutters' as the English called them—of the infamous Free Companies. Irrespective of the fact that the French called them all 'English', these *routiers* were largely Gascons, though they also included Englishmen as well as Bretons, Spaniards and Germans. They were disciplined companies of veteran soldiers who were unwilling to give up their lucrative trade even in peacetime, which was when they became the most troublesome. Peace meant unemployment, and without an employer they lived off the land like armed brigands, with little or no respect for church, property or life. Some sought alternative employment in Spain while others, like Sir John Hawkwood's 'White Company', immortalised in the book of that name by Sir Arthur Conan Doyle, went down into Italy, where such freelances were known as *condottieri*. Indeed, the warring city-states of Italy were then Europe's leading employers of mercenaries, and the very first Free Company of all, Roger de Flor's 'Catalan Grand Company' of 1302, was actually of Italian (or, more accurately, Sicilian) origin.

The English were finally ejected from France in 1453, when the curtain at last came down on the Hundred Years' War. Despite the fact that the longbow had not scored any particularly notable successes since Agincourt it was nevertheless to remain the principal weapon of the English soldier for a further 100 years. Both sides used it during the Wars of the Roses, notably at St Albans in 1455, Towton in 1461, Edgecote in 1469 and Tewkesbury in 1471, inflicting appalling losses on one another. But also in that period were sown the very seeds of the longbow's decline; for at some time between about 1445 and 1470 the handgun was introduced from the continent, and from then on it was only a matter of time before the grey goose shaft was finally ousted by 'villainous saltpetre'.

Firearms

The date of the discovery of gunpowder is not really known, although the mid-13th century monk Roger Bacon is often credited with this dubious honour. What is fairly certain is that the first cannon (and I use the word loosely) appeared in the 1320s, being first recorded in England in 1327 where they were called *crakeys* (*crake* being the name at first applied to gunpowder). These early guns were crude and unreliable and basically consisted of nothing more than an iron tube with a touch-hole to which the match was applied; they fired stone shot, some of the very earliest even firing heavy metal arrows! Evolution was relatively slow, and although Edward III had some sort of field artillery with him as early as the Battle of Crécy, where three guns are recorded, truly mobile artillery which could be easily moved during battle did not appear until the 15th century. The Hussites, Bohemian religious extremists, were using light cannon mounted on wagons in the 1420s, but it was not for about another 50 years that artillery mounted on a wheeled carriage actually became available. The handgun, something of a German speciality, was in general use by about 1385, at first in the form of a small cannon mounted on a wooden stock but before very long with a proper shaped stock, a longer, slimmer barrel and a trigger for applying the match. The handgun was relatively ineffective until about 1450, and although as early as 1420 there were a few artillery pieces available capable of firing stones weighing up to *800* lb, cannon only first played a decisive role in battle at the very end of the 15th century, by which time the Middle Ages had drawn to a close and the age of the Renaissance had begun.

Strangely enough it is this later period, of the longbow and the primitive cannon, that is the most popular with mediaeval wargamers, probably because for many years the only figure ranges available covered either the Hundred Years' War or the Wars of the Roses to the exclusion of all else. This was possibly a result of patriotic figure manufacturers wanting to see England's most famous mediaeval victories recaptured on the wargames table, but more probably it resulted from a dearth of information on the earlier part of the Middle Ages. This situation has recently been rectified by the publication of several books by the Wargames Research Group and Osprey which specifically cover the earlier period, and the manufacturers are now vying with one another for the most comprehensive ranges of figures, based largely on the WRG publications. So whatever your personal fancy may be—whether refighting the Spanish *Reconquista,* battling with the Teutonic Knights in the snowy wastelands of Prussia and Lithuania, or simply smashing the French (again) at Agincourt—there is little doubt that before very long you will be able to find any mediaeval army that you could possibly think of *somewhere* amongst the vast array of figures that are fast becoming available.

Rules for mediaeval wargaming

As for Ancients but with the addition of Super Heavy Knights (SHK), Welsh longbows and primitive gunpowder artillery. SHK move at the same pace as EHC but count as SHC in mêlée. Deduct −1 from all weapons except crossbows and artillery when firing against SHK. Longbows count as crossbows at short range (ie, under 50 paces). Primitive gunpowder cannon count as heavy artillery for movement and firing; if opponents lack artillery, take immediate morale test for each unit as it comes under fire, deducting an extra −2 from score.

Chapter 3

Renaissance and 17th century warfare
by George Gush

A few years ago this period, with the partial exception of the English Civil War, was neglected and little known, but today it has become one of the recognised historical 'divisions' of wargaming, and is probably not far behind the 'top three'—Ancient, Napoleonic, and World War 2—in popularity, having certainly overtaken such old favourites as the American Civil War and the 18th century. This is partly a result of an increase in books and articles about the period, accompanied by a great growth in wargaming equipment for it. There has long been a fair selection of figures for the Civil War, but recently a great range of models, both for this and the rest of the period, has appeared in 1:300, 15 mm and 25 mm scales, produced by some ten makers. There are at least six published sets of rules available, ranging from the very simple to the complex, but having one inestimable boon for the wargamer—commonality in the base-sizes required for mounting the figures, so that the same force can be used under different rules without difficulty. Several of the rules are accompanied by army lists which give a beginner a useful idea of the possible armies, their composition and the classification of different troops under the rules. There is even a club—the Pike and Shot Society—devoted to the military history of the period and wargaming set therein, with a wargames championship and a bi-monthly magazine. And there are several re-enactment societies like the Sealed Knot, in which the participants actually arm themselves with pike and musket and march to battle in 17th century costume.

However, the increase in popularity is certainly also due to the great inherent interest of this period of military history, and the particular advantages and attractions that it offers the wargamer, recognition of which has steadily grown since Midlands wargamer, Dave Millward, started the 'boom' by introducing the period in the 1971 National Wargames Convention.

Historically, this is often referred to as the 'Renaissance' or 'pike-and-shot' period, but might perhaps be better thought of as the 'early gunpowder' age for, although gunpowder had been in use for nearly two centuries before, it was in the 16th and 17th centuries that it actually changed the face of warfare, helping to end the Middle Ages and usher in the modern world. At the beginning of the 16th century, armies and fortifications were still essentially those of the Middle Ages, but by the late 17th century regular paid troops had replaced wandering mercenaries and feudal levies, uniforms had appeared and armour practically vanished, stone castle walls had given way to massive earthworks and, except for a few lingering pikemen, battles were fought with essentially the weapons, formations and tactics that would prevail until the age of Napoleon.

The new armies and their weapons helped create the new strong monarchies, rolled back the Turkish tide from Europe and, with the new ships, began the age of Europe's world domination, while the new firearms affected warfare between non-European powers everywhere from West Africa to Japan. The 16th and 17th centuries saw the peak and decline of the Ottoman Empire, the golden age of Spain, the meteoric rise and fall of Sweden and Poland as great powers, the creation of the Russian Empire and the spread of the seaborne ones of Portugal and Holland, the Moghul conquest of India, the titanic struggles of Valois and Hapsburg, the endless Wars of Religion unleashed by the Reformation and the battle for the balance of power against the might of Louis XIV of France. Hardly a comfortable era into which to be born but, now it is safely over, one offering unlimited scope and interest to the wargamer.

Obviously we have not space here for a full account of the armies and wars of the period, but we can trace the main developments and take a look at some armies typical of particular periods and areas, taking Europe first, then very briefly the non-European world, and finally the British Isles.

European development

In Europe, the first half of the 16th century saw wars of Swiss and Swedish Independence; the fall of an independent Hungary before the Turks at Mohacs in 1526, and the successful defence of Vienna against them in 1529; the revolt of the Comuneros in Spain and the beginning of religious wars in Switzerland and Germany; in the north, Denmark and Sweden fought with the Hanseatic towns, and to the East the rising power of Muscovy struggled with Teutonic Knights, Tatars, Lithuania and the remnants of the Golden Horde. Unquestionably, however, the era was dominated by the 'Italian Wars', some ten successive campaigns between the French Valois rulers and those of Spain and Hapsburg Germany, extending from 1494 to 1559 and from the Low Countries to Naples, and involving on one side or other—often both—the Swiss, the Pope, all the independent states of Italy, the Netherlands, England and even the Turks!

At the beginning of these wars, the fully armoured mediaeval knight with his lance and warhorse was the only standard type of soldier across Europe and still regarded as the chief arbiter of victory; cannon were for besieging castles, and most foot soldiers were fit only for skirmishing with crossbow or clumsy handgun, with a few exceptions like the English longbows and the Swiss, whose great blocks of well-drilled pikemen, often unsupported by missiles, made them the supreme European infantry. It was the French whose invasion of Italy in 1494 gave the initial push into a new era of warfare, and the French army of the early Italian Wars, an interesting mixture of old and new, shows the direction of some of the changes. It still centered around the knight, but in the form of 'Gendarmes d'Ordonnance'—regulars paid and maintained by the King. As was usual, they were organised in 'lances' which included both servants and other combat troops, 'archers' (really lancers, although they might still carry crossbows) who supported the heavier cavalry.

The increasing provision of lighter and cheaper types of cavalry, more suited to scouting and skirmishing, was a feature of the period, and the French army included the most-favoured type, mounted arquebusiers rather like the later dragoons, lightly equipped and able to operate on foot or mounted. At first they would be hired Italians, later Frenchmen. Another type were Albanian

Renaissance and 17th century warfare

The photographs accompanying this chapter show a sequence of moves from an English Civil War skirmish between Malcolm Dove and Clive McLeod, umpired by George Gush. Colonel Dove was in charge of a valuable Parliamentarian convoy which Lord McLeod was determined to intercept . . . In this first scene the convoy is progressing through Little Pottering and the Royalists have not yet appeared.

The Royalist horses have appeared in the foreground and some of Colonel Dove's pikemen prepare to sell themselves dearly in defence of the convoy.

The Parliamentarian horses have now joined in but the musketeers supporting the pikes are in difficulties.

'stradiots' with javelin and shield. The foot centered about hired Swiss pikemen or German 'lansknechts' identically dressed and armed. These were supported with gradually increasing numbers of 'shot' with crossbows or, later, firearms. The wars were to end the short reign of the unsupported pikeman, which finally closed at Bicocca (1522) where Spanish arquebusiers and artillery stopped, and permanently sobered, the dreaded Swiss.

The new matchlock arquebus, the first effective hand firearm, was carried by hired Italian skirmishers, but the French infantry, who were still mercenary bands rather than a national army, were either drilled in imitation of the Swiss, or old-style crossbow skirmishers, their massive 'arbalests', with steel bow drawn by a windlass, shooting up to 300 paces, as far as early guns although with somewhat less power and more strain on the operator. The most modern feature of the French army was the artillery—eight-foot bronze cannon on the new wheeled carriages, which made very short work of high stone walls and were also effective in the field. Fornovo, 1495, was the first battle won by field artillery.

By the end of the Italian Wars infantry firearms were coming to dominate European warfare, although still requiring the support of the pike, and Gendarmes were being supported and supplanted by increasing numbers of firearm cavalry. Guns were becoming standardised, and in a successful effort to combat them the Italians had developed a new system of fortification, with low massive masonry walls behind a ditch, beyond which a shallow sloped bank, the glacis, protected them from siege guns, while providing a 'killing-ground' for the artillery which could now be mounted on the ramparts and for the fire of arquebusiers massed in a 'covered way' on the lip of the ditch. The old round towers were replaced by solid angular bastions or separate ravelins, which could sweep every inch of the approaches with fire. The numbers needed to besiege such defences successfully may have contributed to another change—the growing size of armies: in the wars up to the Peace of Cambrai (1529) no army probably exceeded 30,000 men; in the 1550s Charles V raised 109,000 in Germany and the Netherlands, 24,000 in Lombardy, and more elsewhere, to a

The view from the Parliamentarian side a move later with Colonel Dove's infantry trying to form a barrier while the wagons get away.

total of at least 150,000—numbers scarcely seen in Europe since the Roman Empire.

In the second half of the 16th century, Wars of Religion reached full spate, particularly in France where no less than *nine* civil wars between 1560 and 1598 involved not only Catholics and Huguenots but troops from Switzerland, Germany, Spain and Britain. In the north, two Livonian Wars involved Muscovy, Poland, Denmark and Sweden; the declining powers of Lubeck and the Teutonic order were also involved in Baltic conflicts and Poland and Sweden became involved in dynastic quarrels; to the south Philip II's inheritance of Portugal sparked off war involving both Peninsular powers and England, while the Hapsburgs continued wars on land and sea with the Turks, involving Venice, Transylvania, Wallachia and the Knights of St John.

Above all, however, this was the Golden Age of Spain, whose Conquistadores had won her the riches of the Americas earlier, and the Spanish were most concerned with their '80 Years' War' with the rebellious Dutch, lasting from 1566 to 1648 with one short truce, and involving many foreign troops and actual intervention by France and England (the Armada being a result of the latter). The Dutch ultimately succeeded, and did much to change warfare with their application of the new Italian fortification system to cheap and nearly indestructible earthworks, as well as with the tactical innovations of their great general, Maurice of Nassau, which were the foundation of the later Swedish system. However, the long wars brought the Spanish forces to unprecedented levels of strength, experience and efficiency, and we may take the Spanish army to represent late-16th century European warfare.

The basis of the army was now the infantry, formed in the famous 'Tercios', each up to 3,000 strong. Permanent units with their own distinctive flags, fixed organisation, 'esprit de corps' and even sometimes uniform clothing (one Tercio, all in black, were nicknamed 'The Sextons') were one of the new developments of the period. 40 to 50 per cent of a Tercio would still be pikemen, formed in a deep Swiss-style block, but now serving primarily as a necessary support for the 'shot'. The latter, usually formed about their pikes in a castle-like formation of lines and corner blocks, were mainly armed with the arquebus, by now roughly standardised to about a half-inch bore, and having a useful range not over 200 paces or so. Shooting a rank at a time, then filing to the rear, arquebusiers could manage a fairly steady fire, but at less than a round a minute per man, it was not sufficient to stop a cavalry attack—hence the need for the pikes. At first a few, but later up to half the shot carried the new musket, introduced by the Spanish in the later Italian Wars. Some six feet long and three-quarters of an inch in bore, a musket could pierce even the heavy plate armour of this period and could kill an unprotected man at anything up to 400 yards, but was even slower to load than the arquebus and required a forked rest to support the barrel when firing. Small numbers of infantry intended for guarding colours, assaulting breaches and other special work carried halberds or the small round shield or 'buckler' and sword much used by earlier Spanish infantry.

Cavalry were still vital. They could sometimes win a battle on their own, as at Gemblours, 1578, and usually formed at least a quarter of the Spanish army. A few gendarmes remained, but the Spanish preferred their own speciality, 'lancers' with at least open helmet and breast and back plates, armed with light lance and pistol. They also had plenty of the primarily firearm-equipped cavalry who had become 'standard' in most countries by this time—the aforementioned mounted arquebusiers, unarmoured or with light breastplate, who would skirmish in support of more heavily armoured close-order cavalry such as the German 'Schwarz Reiter', trotting in deep formation to discharge their pistols rank by rank and wheel away to reload in the 'caracole'.

The first half of the 17th century saw continuations of the Hapsburg-Turkish and Dutch-Spanish conflicts and further civil wars—the Huguenot revolt and the Fronde—in France. In the Baltic Denmark fought Sweden; the Poles fought Russia, Sweden, the Transylvanians and the rebel Cossacks of Bogdan Chmielnicki, and in the Peninsula Portugal, with English and French aid, re-established her independence. All these were overshadowed, however, by the gigantic and horribly destructive Thirty Years' War of 1618–48. Fought primarily across the German Empire, there were subsidiary campaigns in France, the Low Countries and Italy, and the contestants included the German and Spanish Hapsburgs, Bohemian rebels, Catholic Princes of south Germany and Protestant ones of the north, Denmark, Sweden, France, Savoy, Holland, England and even Transylvania, although not all at the same time!

The most effective army of the day, representing the latest ideas in organisation, equipment and tactics, was the Swedish army of Gustavus Adolphus. Its cavalry were rather more numerous, proportionately, than those of the earlier Spanish army (there was a marked trend toward more cavalry at this period—possibly because they were better at foraging and getting across ravaged areas without starving!) They were of two types—a few cuirassiers, similar to the earlier Reiters, in heavy three-quarter armour (they were much more numerous in other armies of the day); and the rest who were called 'Arquebusiers' or 'light horse'. Like their earlier namesakes they limited their protective arms to helmet, breast-and-back and buff coat, but unlike them were armed with sword and pistols and trained to charge home at the gallop in three-deep line. The equivalent of the old-style arquebusiers were the small numbers of dragoons, basically trained as mounted infantry, unarmoured and musket-armed.

Things are going badly for Colonel Dove; his cavalry have lost the mêlée and his foremost infantry unit is in flight.

The infantry still consisted of armoured pikemen supporting 'shot' with firearms, but the battlefield units had now become 'squadrons' (called 'battalions' in most other armies) of only 400–500 men, conferring much greater flexibility on the army. They were drawn up in shallower formation, only six ranks deep, partly because the 'shot' were now trained to fire by 'salvoes', up to three ranks firing together. All the shot were armed with muskets which, although still matchlocks, had been considerably lightened and perhaps provided with lighter rests. Additional musketeers were detached from their units (which were supposed to have three musketeers to each two pikemen) to form a skirmish line or 'forlorn hope' before the army, and to form up among and support the cavalry wings. In what was much more of a 'regular' army than earlier ones, drill and discipline were of new standards. The artillery in particular was improved: Swedish gunners were regular soldiers formed into regiments rather than the hired specialists of earlier days, who seldom felt it incumbent upon them to get mixed up in vulgar close-quarter brawls! Gustavus made wide use of light, mobile 3 pdrs which could accompany infantry units and support them with blasts of 'hailshot', while the use of paper cartridges allowed them to fire more rapidly than the musketeers.

After they had demonstrated their superiority to the Spanish system at Breitenfeld (1631), such methods were adopted in most countries so far as problems of recruitment and training allowed.

In the later 17th century, England took a much greater part in European conflicts, joining, under Cromwell, in France's continuing war with Spain, 1648–59, fighting three wars of her own with the Dutch, and being closely involved in the wars brought about by the ambitions of Louis XIV. Of these the War of the League of Augsburg, 1688–97, was the greatest conflict of this last period, involving fighting in the Netherlands, Germany, France, Italy and Ireland, and with France (and the exiled James II of England) facing a vast alliance of England (William III of Orange), the German states, the Emperor, Holland, Savoy and Spain. Another major conflict was the first Northern War, of 1660–1665, involving Sweden, Denmark, Poland, Russia, the Empire,

The crisis! The retreating infantry pour in panic through the ranks of another regiment, and the Cavaliers have broken the Parliamentarian dragoons.

Transylvania and, at sea, even England and Holland; Russia fought a series of wars, of which this was one, with Poland, and began a series against the Turks, who were thrown back from Vienna for the last time in 1683 by John Sobieski of Poland and were subsequently forced back out of Hungary, beginning their long retreat from Europe.

By the latter part of this period European armies had become pretty uniform, so that we can consider all rather than one in particular. All were now regular, semi-standing armies, paid and uniformed. Cavalry, though still important, were no longer quite the battle-winning arm they had been, and tended to form a rather smaller proportion of most armies—30 per cent or less. They still included cuirassiers, at least in French and German armies, with corselet and sometimes armoured gauntlets, although outside Bavaria the old lobster-tail helmet was replaced by a metal-reinforced hat. Dragoons were growing in numbers and tending to become more capable of mounted action, and one or two new types had appeared, such as 'carabineers' armed with a type of blunderbuss firing several musket balls, and the hussars of the French and Austrian armies, derived from the Croat and Hungarian light horse employed in the 30 Years' War, and still essentially irregular skirmishers. Generally the cavalry still carried sword and pistols (now flintlocks rather than the earlier wheellocks) and the usual tactic was a charge home, but at the trot rather than the gallop (except in the Swedish army).

The infantry had been changed much more from earlier patterns and were now clearly the backbone of the armies. Firepower had been increased by the adoption of cartridges and better drill, and the superior flintlock musket had about half-replaced the matchlock by the end of our period. As important was the appearance of the bayonet, making every musketeer his own pikeman. The French began to use bayonets experimentally from the 1640s, and their use gradually became more widespread after the 1660s although, as these were plug bayonets, blocking the musket barrel when fixed, some pikemen were retained; by the 1680s the proportion had fallen to about one man in every five, and

However, the convoy is now nearing the safety of the river and the foreground Parliamentarian infantry are standing firm.

thereafter pikes rapidly disappeared, especially with the adoption of socket bayonets around the end of the century. Although pikemen faded out, a couple of other specialist types made their appearance in this period; firstly the grenadier, an élite assault-trooper for attacks on fortifications, usually distinguished by a fur cap, and armed with grenades and, at first, an axe for hacking down palisades; and secondly the fusilier, armed with a light flintlock musket and specialised to guard and assist the artillery train. In battle, infantry battalions formed up in five or six deep lines, fairly wide-spaced, and relying on firepower rather than shock for their effects, and were still often supported by very light guns attached to the battalion.

Developments outside Europe

Having concluded our survey of European developments, we must take a quick glance at the world outside, although there is not space for more than a mention of the innumerable wars of the period. Besides those Turkish wars with Europeans already mentioned, the Ottoman Empire defeated the Egyptian Mamluks at the beginning of the period, and fought a whole series of wars with the Persians, after the latter had emerged from the wars of the rival Turcoman peoples of the Black Sheep and the White Sheep. The Persians also fought the Moghuls, who conquered North India in 1524-56, under Babur and his successors. Further afield were all sorts of civil and international wars, greatly complicated by the intervention of empire-building Europeans—the Dutch gave the Japanese arquebus and cannon to be employed in their feudal struggles; in China the Jesuits introduced artillery into the Ming-Manchu rivalry, and firearms were used when Russians, advancing across Siberia, were thrust from the Amur Valley by the Manchus; North and South Vietnam employed cannon in their lengthy internicine wars, and the Portuguese introduced firearms into the wars in the Horn of Africa, where they backed the Christian Ethiopians of Emperor Claudius against the Turk-backed Moslem Somalis (some familiar-sounding set ups, weren't there?). Struggles for power along the North African seaboard involved the Turks, Spain, Portugal, local rulers, England and Morocco, which defeated the Portuguese soundly at Alcazar (1578) and used firearms to conquer the West African Songhoi Empire. Across in the New World, stone-age empires fell, in the 1520s and '30s, to the arquebusses and lancers of Spain.

Our example of a non-European army must obviously be that of the Ottoman Turks, by far the strongest and militarily the most advanced non-European power. This was built round a substantial paid regular standing army, long before any such thing had appeared in Europe, consisting of regular artillery, supported by mortar, wagon and ordnance corps; the famous Janissaries, primarily firearm infantry with long arquebusses, but quite prepared, unlike their European equivalents, to get stuck in with their scimitars when necessary; and lastly the six regiments of 'Spahis of the Porte', cavalry armoured in mail and light plate and armed with light lance, mace, shield and composite bow or javelin. In wartime these were supplemented by feudal troops similar to the Spahis, and vast hordes of bow-armed light cavalry and light infantry with bow, spear or firearms, mainly enlisted for loot and fit for little but skirmishing.

Most armies from Egypt and Moscow to India were made up of similar troop types, with local specialities like Moghul war-elephants or Muscovite Streltsi musketeers with their two-handed axes. Wagon-laagers were often employed to protect the usually pikeless infantry of these cavalry-based forces. Apart from the spread of firearms, so that even cavalry often had pistols by the late 17th century, there was relatively little development for us to trace amongst Eastern armies.

Britain

We conclude our survey with a separate look at Britain, 1500–1650, in the 16th century because of our relative isolation from European wars; and in the early 17th because of the particular interest of the English Civil War.

16th century England was indeed a bit peripheral to the great events of Europe—effective at sea, but on land prone to spasmodic and ill-organised expeditions which usually collapsed in disorder and disease once the beer ran out! This, however, meant that English armies were interestingly different from those of the Continent, until the 1580s continuing to depend on the famous longbowmen. The longbow could deliver up to six shots a minute, with light arrows carrying up to perhaps 300 yards, and heavier 'sheaf' arrows to 150-plus, with considerable armour-piercing capacity at close range. However, fatigue, hunger and weather could greatly affect the longbowman's shooting, and despite the efforts of Tudor governments to encourage archery and ban other sports, like football, in its favour, the traditional skills needed were beginning to be lost. Longbows were supported, not by European-style pikemen, but by billmen with weapons rather like the European halberds (probably because the English did not have to face good cavalry against whom pikes were essential). English cavalry were largely 'Border Horse' or 'Javelins', light horsemen with spear and perhaps crossbow or, later, pistol. Armour was less in evidence than overseas, too, being usually confined to a mail shirt or 'jack' of steel plates sewn in a leather jerkin. Their opponents were equally unusual. In the continual wars with Scotland between 1482 and 1550, the Scots fielded dense but ill-supported 'schiltrons' of pikemen rather like the Swiss, and in the incessant warfare in Ireland, which culminated in the great revolt of the 1590s and the intervention

Renaissance and 17th century warfare

The end of the game. The Royalists have succeeded in capturing two wagons but Colonel Dove's inspired leadership (and some lucky dice throws!) have otherwise produced a stalemate.

of Spanish troops, English armies, stumbling into fortifications and obstacles hidden in the woods, were assailed by wild Kern hurling javelins, and Viking-style Gallowglasses with mail shirts and two-handed axes.

English troops fought in foreign armies—notably that of the Netherlands, and English rulers hired foreign mercenaries, so that England gradually drew closer to the Continent in military matters, and by the great Civil Wars of 1642-1650 the only major remaining difference was continued reliance on a local militia system for raising troops in time of need, as opposed to European use of regular or mercenary troops. The Civil War armies reflected up-to-date European practice, with pike and shot battalions deployed in the Swedish or the deeper Dutch order and cavalry mainly in buff coat and breastplate with sword and pistols employing either the Swedish charge at the gallop introduced by Prince Rupert or the more controlled advance at the trot favoured by Parliament. The horse were varied by Parliamentary cuirassiers—the famous 'Lobsters'—and Scots lancers, and supported by small bodies of dragoons. In Scotland Montrose's Royalist army was of quite different type, combining firearms with the traditional bows, two-handed claymores and axes of the Highlanders, almost unstoppable in their usual ferocious charge.

The Civil Wars have particular interest for the wargamer, largely of course because of the abundant information on them available in English, and their natural attraction for players in these islands who may readily visit the scenes of battle and siege. Their relatively small scale, which makes it easier to translate real battles to the table-top, is another attraction, and perhaps also the greatly varying states of morale and training as armies were built up from hot-headed gentry, raw peasants and townsmen or little-trained militia, ultimately to the standards of perhaps the best-trained army of its day, Fairfax's New Model. Moreover, although there were plenty of sieges, the Civil Wars were generally notable for the number of assaults and pitched battles, large and small, which seem to show a stronger spirit of decision than in some conflicts of the period. Perhaps this was because of the number of able leaders these wars produced, among them Prince Rupert of the Rhine, one of the most dashing cavalry leaders of all time, the Marquis of Montrose, who possibly won more battles against heavy odds than any other general of history, and the able soldiers ultimately found by Parliament, Sir Thomas Fairfax and Oliver Cromwell.

Costume

Whether he chooses to refight the English Civil War or to explore the lesser-known conflicts of this era, the wargamer can find endless interest in 16th and 17th century warfare. As even our brief survey indicates, there was an unparalleled variety of armies, troops and weapons—really all the weapons and armour of the Middle Ages, *plus* the disciplined regiments and the firearms of later times, with some completely unique things like Winged Hussars thrown in as well!

For the player concerned with the appearance of his armies (and, as the Introduction suggests, most are) this must also be about the most colourful period of all, particularly in its earlier stages, with people like the Swiss and Lansknechts dressed in the most extraordinary multicoloured costumes, one sleeve one colour and one another; one leg perhaps bare, the other clad in a baggy trouser leg slashed into strips to allow a contrasting undergarment to show through; the doublet likewise but in quite different colours; and the whole tastefully garnished with ribbons and ruffles and topped off with, say, a crimson beret covered with plumes dyed in three different hues! Even these sartorial excesses could be rivalled by the aforementioned Winged Hussars—a mounted Archangel in plate armour gives the general idea—or such exotica as Turkish 'Dellis' clad largely in lion and leopard skins, with the odd eagle's wing tacked on here and there! Even when uniform began to appear, every regiment and often every company had its own coat colour, and in a much greater range than later times, with purple, black, tawny and yellow appearing, for instance, in English Civil War armies. Moreover, flags were not only highly coloured, but usually carried by every *company,* rather than being confined to regiments as in later times. The painter has plenty of creative scope, too, for until very late in the period uniform regulations were an unheard-of invasion of individual liberty!

This lack of uniform was a problem for commanders of the day, who were forced to arrange 'field signs' for identification—thus the Imperial general Wallenstein ordered his men to wear a piece of red cloth 'on paine of lyfe'—but for the wargamer it is actually a help, since it means that practically any unit in, for example, the Civil War period, can serve for either side (the popular idea that Cavaliers were all lace and lovelocks, while Roundheads were distinguished by short haircuts, helmets with faceguards and disapproving expressions, being a fallacy). It is very convenient being able to slot in the available troops on either side as needed, with at most the provision of a spare set of flags, and perhaps, for the purist, officers (since the latter sometimes distinguished themselves as

Royalist or Parliamentary by wearing, respectively, red or orange sashes). Indeed, a few more flags, and your Civil War armies can serve as troops of the Thirty Years' war, who were similarly dressed and armed, or indeed those of other early 17th century wars. In the 16th century too, although there were fewer 'universal' troop types, some, like Gendarmes, could serve in most armies unchanged, and other troop types, as mercenaries, actually *did* serve in many different forces. Thus Lansknechts served in most armies, and could with suitable cantonal flags also act as Swiss; 'Reiter' cavalry were almost universally used, and so on. The wargamer of this period, by starting with such mercenary types, can lay the foundations of quite a few different armies without having to go to the expense of vast numbers of soldiers.

The charm of the period

When we come to the actual playing rather than the modelling side, variety is again one of the great charms of this period. We have already seen the wide variety of weapons in use, and the great differences in protection, formation and mobility that could exist, even between units in the same army. But beyond this, the Renaissance wargamer, like the real generals, has to deal with the widest variations of outlook and training among his miniature followers. Not for him the drilled uniformity of later days; if he uses, for example, the Wargames Research Group rules for the period, which, like most others, seek to reflect this aspect, he may find himself, say, with a core of 'B' or 'C' class troops, steady, and, by the lowish standards of the day, disciplined, but as for the rest . . .

Those Stradiots, for example, typical 'D' class irregulars; they could be set to slip round the enemy flank and harass his rear—but what about the baggage train behind his lines? If they sight *that* there is every chance they will spend the rest of the day looting and not fighting! Then perhaps there are some 'E' class levies, untrained and unenthusiastic. The rules, like most for the period, include a points system which allows us to pick roughly equally matched armies from this very varied material, and the 'E's are cheap—we could afford to lose them—but the inevitable rout could start a panic; better hide them in the wood, perhaps.

The inevitable mercenaries can be a problem too; those well-armoured and mounted Italian condottieri are a case in point—'M' class, skilful but not apt to take risks; if we don't give them plenty of support they won't charge, just sit there looking impressive while other people do the fighting. Quite the opposite is the case with 'A' class cavalry—they could be Gendarmes for example, or Rupert's Cavalier horse. Proud and brave, but they are independent gentlemen rather than what a later age would consider soldiers—*they* will decide for themselves when honour demands a charge, then off they'll go, orders or no orders. An 'A' class Gendarme is just an 'uncontrolled advance' waiting to happen—all the general can do is try and make sure he is pointing at the right enemy when it does! A bit different from 18th century or even Napoleonic generalship! One soon comes to understand the thoughts of the Frenchman, Montluc, advising his commander on the field of Lanzo (1522): 'Our Italians can pair off with their Italians; give Admiral d'Annebault and Captain Ysnard the French companies and put them opposite the Germans; the rest of our French infantry under M. de Bonnivet can take on the Spaniards . . .'

There are also, of course, great differences between *armies* in this period, so that the Austrians in the later 17th century, for example, had both special tactics and special equipment (such as extra cavalry armour and battlefield obstacles) which were reserved for their wars with the Turks and not used in Western Europe. This can be an even bigger problem for the wargamer, who may well, out of interest as to what would happen, or out of necessity, find his army facing foes it never met in real life—English against Samurai, or Montrose against the New Model, say. A few hints for these contests of unlike against unlike may be in order, albeit from one whose own record in command is hardly unblemished. An infantry army against a mainly cavalry one is probably the commonest problem—most European armies against Turks or other Easterners, say. Pike-blocks will give a firm base but initially you will have to stand, making sure above all that your own horse support the infantry and don't get lured away and overwhelmed by the superior enemy cavalry. However, unless you have overwhelming firepower you won't win by standing still, but an alert general who has arranged the right signals and orders may be able to take advantage of that moment, inevitable with cavalry, when most of the enemy horse are off the field, pursuing, or on it, rallying and temporarily *hors de combat,* to advance and overwhelm any foot or guns the opponents have (won't work against Tatars, who haven't any!). Another difficult task is that of early Scots or Swiss—nearly all pikes—against armies much stronger in missile fire. Provided the latter are not all mounted (in which case you can only form ring and wait for nightfall!), the main point is the pretty obvious one—don't hang about getting shot, but act—advance on the enemy in proper Swiss fashion, guarding the flanks of your pike blocks, which are very vulnerable to disorganisation, with any ancilliary halberds, crossbows or whatever that you have.

Generally, in fact, the tactics actually adopted in the period will be found to work best. Thus a Turkish army is best formed with plenty of cheap rubbish at the front to soak up enemy attacks, guns and Janissaries behind to soften up pursuing and disordered enemy units with their fire, and Spahis in reserve on either flank to charge them, while horse-archers attempt to generally outflank and harry the foe. An English Civil War army, or indeed most later European ones, will have a centre of infantry pike blocks flanked by musketeers, with any artillery either in front of, or if there is suitable high ground, behind the army. (Don't expect artillery to win your battles in this period, but it can be quite useful—in particular, if you have more than the enemy, he will probably have to attack you, and you can plan on this assumption).

The cavalry will be drawn up on each flank, in bodies down to 100-strong (five model figures if using WRG rules), and preferably in more than one line, since victory in a cavalry fight usually goes to the commander retaining the last uncommitted unit. A small cavalry reserve behind the centre was common, and can be useful. The task of English Civil War cavalry was to deal with the enemy horse; this achieved, the problem is to stop them vanishing from the field in pursuit of fugitives or loot (much easier with Ironside 'B's than Royalist 'A's); provided they stay, they can then intervene in the central infantry fight, often with decisive results.

Thus, at, for example, Naseby (1645), the last major battle of the First Civil War, the Royalists, outnumbered more than 3:2, actually won the cavalry battle on their right and looked like winning the infantry fight as well, but finally lost the battle—and the war—when Rupert's victorious Cavaliers vanished from the field, while Cromwell's horse on the other flank rallied and attacked the left of the Royalist foot, most of whom were, consequently, captured.

Apart from the set-piece battle, the period in fact offers several other wargaming possibilities. Although, contrary to some accounts, there were a good many general actions—an average of at least one a year throughout the 30 Years' War—they were greatly outnumbered by sieges and assaults on fortifications (four of the major battles of the Italian wars—Ravenna, Marignano, Bicocca and Pavia—were in fact assaults on entrenched camps). These have their own interest and could involve many intriguing weapons—petards, mortars, mediaeval-style catapults, the incendiary sticky-bombs used by the Turks at Malta and the firepikes which panicked the Roundhead defenders of Bristol. The new fortifications called out a new form of siege warfare finally perfected in the late 17th century by Louis XIV's great engineer, Vauban, his system of saps, parallels and batteries ushering in the formal siege of 18th century style. Sieges and assaults are rather neglected by wargamers, but can provide most interesting games, and while it is rather difficult to get a Vauban fortress on to the wargames table, the improvised earth star-forts and 'sconces' of the English Civil War are a different matter, and the many country house sieges of that war provide particularly interesting small-scale actions, ideally suited to the wargame. WRG rules include a section on sieges and assaults, and others may be readily adapted.

The incessant Mediterranean galley warfare of these centuries also has great possibilities; though we are not here concerned with pure naval warfare, these were really amphibious struggles in which ordinary infantry weapons, and fortified places on land, could often determine the outcome of a sea-fight. Suitable small-scale galley models and rules are both available.

Whatever type of action one chooses, the 16th and 17th centuries have much to offer—battles small enough to be reasonably reduced to wargames dimensions, formal deployments and formations which look most realistic on the table-top, and the colour and variety already mentioned. Despite the growing power of firearms, the cavalry maintained their importance as the chief arm of decision almost to the end of the period, and this, with relatively limited firepower, tends to give brisk and decisive games.

Finally there is the characteristic feature of the period, which distinguishes it from others, the fascinating problem of integrating the actions of pikes, shot and horse—rather like the old game of scissors-paper-stone—the pike can stop horse but are largely helpless if unsupported in the face of shot, the shot cannot stand against horse, horse cannot break pikes, and the wargames general must struggle to combine these arms so as to bring him success. We may leave him with the advice given by Robert Ward, just before the Civil War, in remarkably appropriate terms: 'I may well compare a pitcht Battell to a game at Irish: each Gamester must have a special eye, not only of his own, but of the adverse Tables, what hits may be given and how to bring home your own men in safety . . . for as in this game there are two principall things which attend the winner, viz. Cunning in handling his Dice and Judgement in placing his men. For I must compare Shot to the dice which . . . is for the most part a principall cause of victory (yet) . . . the Shot of themselves are too weak for to resist the Horse, unless the wisdome of the Generall place them in such places of advantage in which they might secure themselves.'

Rules for 'pike and shot' wargaming

As George has related in the preceding chapter, the principal development affecting warfare in the 16th and 17th centuries was the handgun—the arquebus, the matchlock musket and, later, the flintlock. In addition, artillery development had proceeded apace leading to a wide variety of more or less mobile field pieces. However, due to the slow rate of fire of both handguns and artillery pieces, gunners of all types were extremely vulnerable to cavalry and required the protection of pikemen. When menaced by cavalry, the pikemen formed a circle—called a 'schiltron'—around the officers and Colours, the musketeers kneeling under the pikes for protection. It takes a full move for a wargames unit to change into or out of this formation.

In this period, pikemen equipped with body armour—morion, breast plate and tassets—count as HI, unarmoured pikemen and musketeers as MI (except when the latter are operating in skirmish order as a 'forlorn hope', when they class as LI). Fully armoured cuirassiers—'lobsters'—are EHC, other armoured cavalry are HC, dragoons are MC and hussars, etc, are LC.

Pistols count as hand-hurled weapons and have a maximum range of 25 paces; muskets count the same as crossbows but have a range of only 120 paces. Musketeers may move *or* fire in a single move.

Artillery for convenience is classified light (up to 6 lb shot), medium (up to 24 lb shot) and heavy (anything over 24 lb). Each model cannon represents two actual guns and is crewed by two figures in addition to the drivers of the horse- or ox-drawn limbers. Light guns may be moved at up to 80 paces per minute and need half a move to be limbered or unlimbered; medium guns can be moved at up to 60 paces a minute and require a full move to be limbered or unlimbered; heavy guns can only be moved at 40 paces a minute and require a full two moves to limber or unlimber. In addition, it takes half a move to traverse a gun through more than 30 degrees to bear on a new target.

The following table gives the 'fire factors' for each type depending upon the target.

Weapon	v:	EHC, HC MC	LC	HI MI	LI Art
Light gun		2	1	3	1
Medium gun		3	2	4	2
Heavy gun		4	3	5	3

These factors are influenced by range. Light guns have a short range of 150 paces, a medium range of 300 and a long range of 450; medium guns have a short range of 250 paces, a medium range of 500 and a long range of 750; heavy guns have a short range of 400 paces, a medium range of 800 and a long range of 1,200. At short ranges add +1 to the above factors and at long ranges deduct −1.

All gunners count as regulars so throw one average dice for each gun firing and refer to the casualty chart on page 40. The majority of infantry throughout this period were regulars of one sort or another, whether mercenaries or trained bands, but the cavalry was notoriously uncontrollable so should count as irregular when taking a morale test.

Chapter 4

18th century warfare
by Stuart Asquith

Before beginning a detailed study of the battles of a given period it is beneficial to have an understanding of the nature of the warfare. In the 18th century—at least the greater part of it—campaigns were limited to the period April to October in any year. This stricture was basically enforced on generals by the need for fodder for the cavalry mounts. The roads over which the armies moved were also of a dubious nature and winter rains generally rendered them unusable—short daylight times also were prohibitive to winter campaigning.

The 18th century was the period of small very professional armies, rigidly disciplined and highly trained. Such bodies were expensive to produce both in the monetary sense and in the time taken to train the raw recruit and mould him into the soulless automaton prerequisite for the battlefield.

Soldiers then were an important commodity, not to be wasted if such an eventuality could be averted. However, it was of no use foraging, living off the land, etc, if it was your own land you were on, so conflicts were inevitable.

Battles were not really the chief aim of the armies' 'top brass'. They were really only considered if no other course was available or if it seemed a fair assumption that one army could—with little loss to itself—very likely put an enemy out of commission. Battles thus became a set of manoeuvres designed to outflank the enemy, who having read the same books was trying to do exactly the same thing. Manoeuvring also sums up the approach to 18th century warfare at a tactical as well as strategical level. Frequent pauses were made in a regiment's advance to check 'dressing', ie, correct alignment of ranks. Such perfections were encouraged by strict discipline and liberal usage of the non-business end of the NCOs' halberds.

Battles may have been infrequent but when they happened they were bloody affairs indeed. Units suffered appalling losses—the six British infantry regiments at the battle of Minden (1759), for example, suffered a combined 30 per cent casualties and yet were the victors. Only rarely, however, did battles result in mêlées—hand-to-hand combats—and wargamers should take careful note of this. Generally the sheer firepower of opposing units was sufficient to settle the issue, being maintained until one side gave under the pressure.

The main weapon development of the period must surely be the arrival of the bayonet. As George has explained in the preceding chapter, this essentially meant that the fire element of the infantry regiment, the musketeer, was no longer dependent on the pikemen for his protection against cavalry. Likewise the faster-moving musketeers were not hampered by the slow-moving encumbered pikemen. Seldom is enough emphasis placed on the introduction of the bayonet and yet it was this single development which dictated the tactics for nearly two hundred years.

The period saw also the development of new styles of troops, such as light infantry, light dragoons and horse artillery. Where relevant below these have been discussed, but in such a brief run down any 'off-centre' types must necessarily receive scant attention.

The 18th century was also the age of the generals, the Duke of Marlborough, Frederick the Great, Eugene of Savoy, Charles XI of Sweden and General Wolfe—all names known to anyone with even a smattering of military knowledge.

Finally the century must also be regarded as a period of sieges. Between the years 1700 and 1745 no less than 46 major sieges took place, the Duke of Marlborough himself presiding at nine of these, for example. The name of Sebastian le Prestre de Vauban (1633-1707), which has surely become so inseparable from fortification, must also be mentioned here since he was responsible for the construction of many of the fortresses which gave the 18th century commanders so many problems.

In the following notes an attempt has been made to concentrate on the more 'popular' periods within the 18th century in view of figures, reference books, etc. Potential recruits to the period, however, should not be over-influenced by this. Many side-shows or small wars, contemporary with the major conflicts, offer tremendous stimulation for wargame re-enactments. The Great Northern War, the Russo-Turkish War (1710-12), the 1715 and subsequent 1745 Scottish Rebellion are all offered as potentially fruitful grounds for the wargamer.

The campaigns

The 18th century say many wars, some localised and some widespread. It will be of use to the wargamer to examine here the potential theatres of war which existed during this period.

Great Northern War (1700-1721)—In the opening conflict of the 18th century Sweden ranged herself against the combined might of Denmark, Norway, Prussia, Poland-Saxony and Russia. There were several major actions such as Poltava (1709) and Gadesbuch (1712) but in the end Swedish dominance of the Baltic was ended and Russia began to become involved in European affairs.

The War of Spanish Succession (1702-1714)—Possibly one of the better-known wars due to the military exploits of John Churchill, Duke of Marlborough, was one of three 'Succession' wars which took place during the 18th century, concerning in this instance the claimancy of the Spanish throne by Philip of Anjou which in the event he retained. The major battles of Blenheim (1704), Ramilles (1706), Oudenarde (1708) and Malplaquet (1709) are familiar to most military students. All were major victories for the Duke of Marlborough, but due to his allied generals' lack of vision the enterprises came to nought and the Duke was relieved of his command in 1711 for political reasons.

The War of Polish Succession (1733-35)—A much smaller affair geographically, this war brought Russia and Austria into conflict with France, Sardinia and Spain, when two brothers contested the Polish throne. Augustus II won the affair and his brother Stanislaus received the Duchy of Lorraine as compensation.

The War of Austrian Succession (1740-48)—This war was a major event and involved most of Europe, Sweden and Russia. The major name of this conflict was Frederick the Great who is synonomous with the strict discipline and rigid drilling of the Prussian army, which in turn typifies the military style of the period. The prime cause was that a Karl Albrecht of Bavaria laid claim to the Habsburg title since the family's male line had ceased, although Maria Theresa of Austria had for a long time been regarded as heiress to the throne. It took eight years of warfare to establish Maria's claim and saw the start of the rise of Prussia as a military power. The major battles were Mollwitz (1741) and Dettingen (1743).

The '45 Rebellion (1745-46)—In 1745 the 'Young Pretender', Prince Charles Edward Stuart, answered the call from his discontented Highland clansmen to come back to Scotland from France to lead them to independence from England. Initially the Prince was received with much enthusiasm but after a series of actions, including Prestonpans (1745) and Falkirk (1746), the Highland army suffered a shattering defeat at Culloden (1746) and the Prince returned to exile.

The Seven Years' War (1756-63)—The Seven Years' War, so called due to its seven-year duration, was another major conflict. Prussia, Hanover and England took on the combined might of Austria, France, Russia, Saxony and Sweden in the cause of securing Frederick the Great's position which he had begun to build after the War of Austrian Succession. The major battles were Rossbach and Leuthen (both 1757), Kunersdorff (1759) and Torgau (1760). The major results were the definite establishment of Prussia as a great power and a large Colonial empire for England. Using the war as an excuse, England took the opportunity of attacking French colonies in India and North America with great success.

Polish struggle for national independence (1768-72)—Once more the major powers got themselves entangled in Polish affairs as they had in 1733. The new King of Poland proved to be merely a Russian 'puppet' and so, to stem Russian dominance, Polish nobles resorted to force and organised an army which was destroyed by the Russians. Prussia and Austria half-heartedly joined the fray in 1769 to prevent a total Russian takeover but in the end Poland was partitioned to prevent widespread warfare.

The American War of Independence (1776-1783)—The War of Independence is probably the most familiar and best documented of all the wars of the 18th century. Long back-dated grudges between England and some of the Colonists flared into open warfare and the conflict saw the emergence of the first American President, George Washington. The war lasted seven years with France coming in on the side of the Americans later on. A point often missed is that many of the Colonists were not at odds with England and many loyalist units fought against the rebels. The final outcome, the Treaty of Paris (1782) recognised the independence of the United States.

Russo-Swedish War (1788-1790)—This somewhat 'odd' small war was begun by the Swedish King Gustavus III to take his public's mind off domestic upsets and worry about the Russian 'ogre'. Little came of the affray and when the dust settled the situation was much the same as it had been before.

The French Revolution (1792)—Without pre-empting Bruce's chapter on Napoleonic warfare, mention should be made of the French Revolution which had its beginnings in the closing years of the 18th century. The huge social gap which existed in France resulted in the famous storming of the Bastille—a Paris

A French army advances on an Anglo-Prussian force during a Seven Years' War battle.

prison—in July 1789. By 1792 the 'Republican' French armies were taking on the Prussians and the Austrians whom they defeated at Valmy and Jemappes. This resulted in 1793 in the first coalition—England, Austria, Prussia, the Netherlands, Spain, Portugal and various German states all aligning themselves against France. This ponderous alliance was countered brilliantly by the French and 1794 saw the Austrians soundly beaten once more at Fleurus, the Netherlands conquered by France with Prussia, Spain and the Netherlands sueing for peace in 1795. The French campaign in northern Italy saw the rise of Napoleon, then 26, and the famous battles of Lodi, Arcola (1796) and Rivoli (1797). The whole of northern Italy came under French control in 1797 with Austria and Portugal pulling out of the war. To strike at his last coalition opponent, England, Napoleon went to Egypt in 1798 but abandoned his army the following year and returned to Paris to consolidate his political position. The turn of the century was marked by the battle of Marengo in 1800 where Napoleon soundly beat the Austrians and set the scene for even greater victories.

These then are the major wars of the 18th century and they serve to underline to the would-be wargamer just how much military activity was going on in the period and how widespread the areas of conflict were.

Still more theatres present themselves as each major war is examined. The Seven Years' War, for instance, sees British and French troops fighting in Canada, where Generals Wolfe and Montcalm fought at Quebec (1759). The battle of Plassey (1757), fought to avenge the fall of Calcutta, made the English the masters of all Bengal as Clive of India with only 3,000 troops defeated the Nawab's army of 50,000.

The choice of the conflict to be re-enacted will be largely a personal whim of the wargamer, but a word of caution here, not all the wars and resultant troop types are covered by figure manufacturers. A wargamer selecting one of the more 'offbeat' wars to recreate may be faced with the problem of converting every single soldier from a nearly suitable commercial casting, which is an arduous task. A unique army results certainly, but the work involved and the time taken to produce it are considerable. Generally the 'popular' periods from the 18th century are the War of Spanish Succession, often called the Marlburian Wars, the Seven Years' War and the American War of Independence. War-

18th century warfare

A battalion of Potsdam Grenadiers, 1756, in 20 mm scale.

games figures for these periods abound and all arms of the services are represented by at least one manufacturer, usually several.

The availability of basic information is a factor to be considered. There is a wealth of text in most libraries on the American Revolution and Marlborough's wars, with slightly less but still a fair amount on the Seven Years' War. Really the whole thing works as a spiral. The publishing of easily read colourful books causes a demand for figures which the manufacturers (generally) happily fulfill. This 'period popularity' sparks off publications of more books and so the process is continued.

Organisation

When one is asked to review an entire century of warfare, the problem of the differing organisations of the various countries means that either a highly detailed coverage or merely a summary results. Whilst it is really only feasible to give general indications and guidelines when discussing such a wide topic, it is interesting to examine some of the regimental structures. It would be tedious in the extreme to detail all the varying structural changes which took place during the 18th century, but a look at the more 'popular' wargame periods within the timespan will serve to highlight the various establishments and changes thereto.

Generally speaking, every army except the British had two or more infantry battalions in a regiment. In the British army this was the exception rather than the rule and only such units as the Foot guards and uniquely the 1st of Foot had more than one battalion. At the time of Marlborough's wars a British regiment of infantry had the usual single battalion, 13 companies strong, totalling just over 1,000 officers and men. Twelve of these companies were variously termed ordinary or 'hat' companies and the 13th was the grenadier company.

Grenadiers had appeared around 1680 and were from the start picked men. Whilst the actual use of the grenade was in vogue, a special headress had to be evolved for these troops in order not to impede the throwing of the grenade. As the use of the grenade gradually diminished the grenadiers retained their distinctive 'mitre' caps and remained the élite company of the battalion. It was because the other companies did not have the special mitre caps that they were termed 'hat men'.

French infantry battalions also had 13 companies with 11 ordinary companies, a grenadier and possibly a 'picquet' company. Authorities are divided on the latter as to whether or not they existed during the Marlburian wars, but it was the picquet companies that were in fact the forerunners of the invaluable light companies later in the century. French battalions were slightly weaker than their English counterparts at 690 men apiece, but generally French regiments had between two and four such battalions. As a last example a United Provinces regiment of the period would have typically two battalions, each having one grenadier and 12 centre companies giving an overall regimental strength of just over 3,000.

The Seven Years' War saw the British battalions slightly weaker at 700, with one grenadier and nine centre companies. The average French regiment, however, had two battalions and fielded over 1,100 men split into a grenadier and a picquet company, plus ten centre companies. Austrian regiments of the period had over 3,000 men, organised into two battalions, each of one grenadier and eight fusilier and centre companies. Frederick the Great's infantry were organised into two-battalion regiments, each battalion consisting of four companies of musketeers and one of grenadiers. The War of Independence saw the emergence, not unnaturally, of the American army. An American regiment—usually called a 'Continental' regiment—had eight centre companies and totalled nearly 700 men. These regiments did not have flank companies.

Their opponents the British had weaker regiments totalling some 500 men organised into eight centre companies, one grenadier and one light company. The French regiments which landed to assist the Americans had two battalions, the first consisting of one grenadier and four centre companies and the second, one light company and also four centre companies. The German mercenary troops sent to help the British, termed 'Hessians' since some of them came from the state of Hesse-Kassel, had four centre companies plus one grenadier company and totalled 700 men.

Mention should be made of the usual practice of detaching the grenadier companies to form combined grenadier formations. This was also done to the light companies as they came into existence. There were both advantages and disadvantages to this system. It gave the Army commander an élite reserve which he could rely on to perform well in battle. Frequently too the grenadier formations were detailed for specific attacks. The famous storming of the Schellenberg Heights in 1704 was spearheaded by grenadiers. The British units which marched to Lexington and Concord at the outset of the War of Independence were the flank companies of ten regiments, as was the main assault force at Bunker Hill two months later.

During the Seven Years' War in Canada, General Wolfe had a formation called 'The Louisberg Grenadiers' which was formed by detaching the grenadier companies from his infantry regiments.

It is not until the later part of the period under review that light infantry emerge as properly organised elements of the infantry battalions. As noted above, these troops too were detailed for specific roles when on campaign—notably so in the War of Independence. To return to Wolfe in Canada briefly, the Louisberg Grenadiers were balanced by a similar light infantry formation under Lord Howe, the unit taking the officer's name.

The system of detaching the grenadier and light companies from their parent regiments was not popular with all officers. Colonels felt the action robbed

Austrian infantry, 1756, in 25 mm scale.

them of their best men, and naturally reduced the battalion firepower by two companies. However, the system continued to be used and was frequently employed during the Napoleonic Wars.

British cavalry or 'horse' regiments during the Marlburian wars had a full strength of over 500 troopers, divided into nine troops. In action, however, the regiment usually formed into three squadrons, with three troops per squadron.

Dragoons were at this time considered to be mounted infantry and were used in a variety of roles. Their organisation was similar to that of the cavalry but slightly less in number, giving some 400 men per regiment. The French cavalry had 12 companies averaging 36 troopers each, with four companies to a squadron and usually four squadrons to the regiment.

The Seven Years' War saw the proving of the Prussian cavalry as a fighting arm along with their infantry colleagues. Prussian cavalry regiments had five squadrons and totalled in the region of 800 men. British cavalry regiments were weaker at approximately 630 and Austrian units considerably stronger at 1,100 men. Only the wealthiest and most patient wargamer should attempt the Austrian army! Russian cavalry regiments generally had five squadrons but the French had just two, each totalling 140 troopers and eight officers.

Little part was played by cavalry in the War of Independence due to the difficult nature of the terrain, but the two British regiments involved were organised into six 37-man troops. Later in the war American cavalry units appeared and were almost twice as strong as the British, with six troops each 68 men strong.

Throughout the 18th century the artillery was the victim of complex establishment arrangements, irrespective of which country one is referring to. Generally speaking the artillery was not under the control of the War Office or its equivalent, but directly under the Ordnance Board which did not have to answer to the army supremos. The arrangement often worked well for the gunners—better pay, better clothing, etc,—but was the bane of the commanding general's life. However, such politics and their eventual solution are outside the scope of this chapter. Suffice to say that Marlborough was well served by his artillery and often had as many as 120 guns for his battles.

In the early part of the century artillery concentrations—or trains as they were termed—were disbanded at the end of a campaign and thus had to be reformed for the next. This was alleviated in the British army in 1716 when the Royal Regiment of Artillery formally came into being. The French finally organised their artillery hierarchy in 1721.

It is difficult to give organisational details for the artillery for, whilst they certainly did have their own company and regimental set up, the 'penny-packet' *ad hoc* use of batteries renders such organisational details purely theoretical.

A brief insight then into the various organisations of the armies. Such troops as hussars, light dragoons and frontier units I have avoided, keeping to the main types of troop in each army.

Detailed organisation charts and Orders of Battle, etc, are all fairly easily available for the major actions of the 18th century. Careful study by the wargamer should provide all the requisite information for the organisation of miniature armies.

Weapons

The main weapon of the 18th century soldier was the smoothbore muzzle loading musket. James II, on his accession to the English throne, expanded his Army and a new scale of firearms was introduced. Generally the muskets were by 1700 flintlocks, the old matchlocks of Civil War fame having been replaced.

The universal adoption through the British army of flintlocks meant there was a tremendous demand for flints and a government factory was established at Brandon, Suffolk, in 1686. This establishment produced all the flints for the Army during the rest of the time flintlock firearms were in general use. As an aside and an item of interest to wargame 'logistic' experts, flints were generally of use for only 20 shots or so before they had to be replaced.

It was in the 18th century that the musket which was to become perhaps the most famous of all was created—the Brown Bess. Whilst its actual date of manufacture is uncertain, the Brown Bess began life early in the period. The musket had a calibre of approximately three-quarters of an inch and fired lead bullets that weighed 14 to the pound, ie, a shade over an ounce (25 grams) each. The cartridge used was a tube of stout paper sealed at both ends and containing the bullet and powder. The firer bit off the rear end of the cartridge, poured a small portion into the musket's firing pan and tipped the rest down the barrel. The bullet was then rammed home with the wooden brass-tipped ramrod with the empty cartridge paper being packed down as well to stop the bullet falling out. The trained soldier using this apparently cumbersome procedure could fire about two to three rounds per minute. The accurate range of the Brown Bess was in the order of 75 yards, its effective range about 100 and maximum 2-300.

The firearms of the other armies of the 18th century were basically similar to the Brown Bess. In the War of Independence the American or 'Continental' infantry were equipped with the French Charleville musket of 1763, which was lighter and more sturdy than the English weapon. Its rate of fire and effective ranges, however, were similar.

Prussian troops in the first quarter of the 18th century were armed with the Henoulgewehr musket made in Liege by Henaul, grand armourer to the King of Prussia, as he was termed. This musket again had similar performance figures to the Brown Bess and Charleville types. Its firepower was enhanced by the drilling of Frederick the Great's infantry who were trained to fire every 16 seconds. Basically then the firearms of the combatant nations of the 18th century were not dissimilar and there was little to choose between them.

Such oddities as the rifle, typified by the Ferguson Rifle which was in limited use by the British in the War of Independence, had appeared on the scene. These, however, remained small in numbers and were the province of the 'new fangled' light infantry units which began to appear in most armies towards the end of the century. Certainly the main infantry weapon for the entire period was the smoothbore flintlock musket and it was on this weapon and its proven capabilities that the generals based their plans.

Pikemen finally disappeared from the British Army in 1702 and after this time all infantrymen were armed with the musket and a bayonet. Initially the bayonet was of a 'plug-in' type and literally plugged into the musket barrel. As a result, needless to say, the musket could not be fired with the bayonet in position. This was soon replaced with the socket bayonet which meant that the weapon could still be fired with the bayonet fixed. As a point of interest the triangular section socket bayonet was so successful it was a standard Army weapon until the end of the 19th century.

The Duke of Marlborough was a great advocate of cavalry using only the sword as a weapon. He allowed his troopers only three rounds for their issue firearms, to be used for guarding the horses in camp only, never on the field of battle.

There seems to have been a great deal of variety in the swords issued to the cavalry so only general comments can be made. The blades were usually three feet long, heavy, basically straight—although some curved slightly towards the tip—with barred hand guards over the hilt.

French cavalrymen preferred to use the pistol and as a result were frequently at a disadvantage against the 'cold steel' tactics of their British counterparts. A typical generalised cavalry pistol would have a nine-inch barrel, fire bullets which weighed 34 to the pound and be accurate up to 25 yards.

Most cavalrymen of the various nations were also armed with differing types of carbine. These weapons were around 43 inches long and fired a bullet of a similar weight to that of the musket. They would also have a bayonet. The so called 'light' dragoons during the Seven Years' War were armed with a sword, a pair of pistols and a carbine.

Ideas on what calibre artillery pieces to have and how to utilise them varied from army to army in the 18th century. Frederick the Great, for example, had scant regard for artillery initially. Then, impressed by the effectiveness of the Austrian guns, he had a change of heart and attached two 6 pdr guns and a 7 pdr howitzer to every battalion in his infantry front line. Heavier calibre guns were deployed at brigade level and generally classed as 'reserve' artillery.

One important note here is that Frederick the Great is generally regarded as the father of horse artillery. Whilst others before him had 'dabbled'—Prince Rupert in the English Civil War had 'flying leather guns'—Frederick organised the artillery on a regular basis, using light calibre pieces.

The French had two great exponents and theorists of artillery in this period, namely La Valliere and Gribeauval, the latter designing his own system of artillery from which Napoleon was to reap some of the rewards.

Generally the artillery of the period can be split into two main types. The small light calibre—say 2 or 3 pdr guns—accompanying the infantry battalions in the line and the heavier piece deployed in a (hopefully) strategic position to the rear. All the guns were smoothbore muzzle-loaders and had accurate ranges in the order of 3-400 yards, effective ranges of 6-700 and maximum ranges of perhaps 8-900.

Tactics

It is extremely difficult to be specific on the complex subject of infantry tactics. These differed from army to army and indeed between regiments within the same army. Basically tactics were left to the discretion of the regimental commanders so as a result many interpretations of the five basic tactical manoeuvres were to be seen. It will suffice within the context of this section to list these basic ideas and then to make a few remarks about the tactics employed during the 18th century, concentrating on the three periods most popular with wargamers, firstly covering the infantry and then the cavalry.

The basic tactics employed were: 1) the advance to virtual contact with a halting/firing movement; 2) static exchange of fire, usually conducted at distances of 60 yards or less; 3) once the exchange was over, the advance to mêlée; 4) defence against attacking cavalry, the square; 5) defence of a prepared position, ie, an earthwork.

It is a well established dictum that weapons dictate tactics. This was certainly applicable in the War of Spanish Succession. With the final disappearance of the pike a few years previously, the musketeers had their own defence against cavalry in the form of a bayonet and did not have to rely on the pikes for this role as they had been forced to before.

As something of a left-over from musket and pike days, the French still made use of deep formations. Officially there were five ranks, but generally troops deployed in three or four lines. This still meant that only the first three lines could fire, the front rank kneeling, the second rank crouching and the third standing. The fourth and/or fifth ranks if they were there presumably stood by and made notes for they could take no active part. This meant of course that a maximum of 40 per cent of the battalion was unable to fire so the maximum fire potential of a unit was never realised. Firing was generally organised by the regimental or battalion commander with the result that there was little or no synchronisation of fire and many misses. Lastly, and most importantly perhaps, there was no way that a constant fire could be kept up since after the volley the men would be feverishly reloading for the next round. It is interesting to note that this system was not peculiar to the French as many German states, along with the Austrian and Russians, used similar methods. Whilst the system was due mainly to the conservative attitude of French officers, it should not really be ridiculed by 20th century thinking. Army officers are notoriously reluctant even nowadays to accept new thinking and stratagems and this was equally true in the early 18th century.

Differing ideas about firing methods were, however, being taught to the soldiery of Sweden, the United Provinces and Great Britain. Platoon firing was the name of the game, with many national, army and regimental variations infringing on the basic idea. When a British infantry battalion drew up on the battlefield it was in line, in a certain order of companies. The grenadier company was split in two and each half stationed on the extreme flanks. The Colonel's and Lieutenant-Colonel's companies would be in the centre with five

18th century warfare

companies on either side of them. The 12 companies—with the exclusion of the grenadiers—were divided into four 'Grand Divisions'. Each of these divisions was then in turn sub-divided into four equal platoons of about 40 men. So in a 13-company battalion there were now 18 equal-sized platoons—16 of them in the four Grand Divisions and one each of the grenadier detachments on the flanks. It should be remembered here that this type of organisation was used only for firing. The 18 platoons then were sub-divided once more into three 'Firing', each consisting of six platoons. The six platoons which made up each 'firing' were interspersed down the battalion line in order to obtain a continuity of fire. The sequence of firing was then as follows:

	First firing	**Second firing**	**Third firing**
a)	Fire	Ready	Load
b)	Load	Fire	Ready
c)	Ready	Load	Fire

The process was then repeated as often as was necessary to convince the enemy of the inadvisability of his attack. Needless to say, whilst this system was effective enough, it was tiring and called for good training and nimble reactions. More accurate fire as a result of closer control naturally resulted, and it also meant that at any one time a British battalion—or anyone else employing the same method—always had one third of its men with loaded muskets in the event of having to deal with any unforeseen enemy action.

Forming square was the basic defence against attacking cavalry and in this case each of the 'Grand Divisions' formed the face of the square and fired as such. Usually the Grenadier companies moved into the middle of the square to defend the officers and colours who were stationed there.

When it came to the defence of earthworks and the like, a six-deep formation was usually employed. The front rank fired, turned and marched to the rear to be replaced on the firing step by the second rank who then repeated the process. Such a technique could only be used behind the comparative safety of defence works.

Several other countries adopted their own ideas. Swedish troops, for example, used a four rank formation. At a nerve-wracking 40 yards the rear two ranks fired over the heads of the crouching front rank men. The unit advanced and then at a mere 20 yards the two front ranks fired and the whole unit closed with the enemy. Such tactics proved too much for the Russians at Narva (1700) and nearly so at Poltava (1709) during the Great Northern War.

By the time of the Seven Years' War a new military power—Prussia—was on the scene, her infantry taking everything literally by storm. Sharp execution of drill movement and rapid fire by the Prussian infantry astounded the Austrians at Mollwitz (1714) and cost them the battle.

In action each Prussian battalion—there being two per regiment—divided into two wings, each of which was further divided into two divisions, each of two platoons. Firing was carried out by individual platoons, left to right along the line. As a final note on their superiority, Prussian infantry were also trained to fire as they closed on the enemy at 75 paces a minute.

Hand-to-hand combat became more and more of a rarity—long before the opposing factions could reach close quarters one almost invariably broke before the sustained fire of the other. The volume of fire and the mass of bayonets generally dissuaded cavalry from attacking and we find the instances of infantry forming square as a defence during this period very few and far between. Only

A patrol of French light dragoons, 1756, in 25 mm scale.

when the cavalry caught an undefended flank, as at Rossbach (1757), were they really effective. Wargame rules should reflect this important facet of the Seven Years' War. Generally the high degree of Prussian training was reflected in the other combatants of the Seven Years' War. As in most cases of military fashion, be it dress or drill, those of the dominant nation of the time are copied, and so it was with the Prussians.

Tactics in the American War of Independence were less dictated by weapons than by terrain. The largely virgin forests and uncultivated lands of North America proved to be no place for the parade-ground precision drill of the European battlefield.

It was in this conflict that the light infantryman proved himself to be the most valuable type of soldier. In set piece actions light infantrymen often formed assault columns, skirmishing ahead of the main body of the army, and generally turned out to be very versatile troops.

The British were hampered by their parade-ground tactics and thought it most unsporting that the American militia and riflemen should snipe at British columns from behind cover. The officers had difficulty adapting to the new tactics forced on them by the terrain and the Americans, and frequently it was only the discipline and firepower of the British infantryman which saved the day.

Gradually a guerrilla type of war occurred, mainly conducted by the American forces with no cumbersome military traditions to uphold. When the British had a chance to fight in European style, the Americans were devastated, as at Brandywine and Germantown (1777). The bayonet was a decisive weapon and won several victories for the British and their German allies.

If the infantry tactics of Marlborough were fragmented then those of the cavalry were even more so in that there was no higher formation than the

18th century warfare

regiment. Larger groups of cavalry were simply assembled for a campaign and then disbanded at its conclusion. Cavalry was, however, still the most effective weapon in open country.

Generally the French troops moved to the attack on a troop frontage. Moving forward at a slow trot to within small-arms range, they halted and fired. In this way they lost virtually all the advantages of shock action.

Marlborough, however, trained his cavalry in the use of the sword, pistol ammunition being issued only in order to defend horses when feeding. Advancing at a fast trot in two or three ranks on a squadron frontage, the British troopers did not pause to fire but crashed straight into the enemy's ranks.

Marlborough usually supported his cavalry with infantry and artillery and was thus able to take full advantage of any tactical gain made by his troopers. It is interesting to note that the full-pelt charge was frowned upon, it being felt more important to maintain the order and closeness of the formation as the troopers charged. One suspects, however, that in the final part of the 'charge' the cavalry would unavoidably pick up speed and hit the enemy with considerable force.

Charles XII of Sweden added a couple of useful paragraphs to the cavalry manual. He advocated an arrowhead formation for attacking cavalry to adopt, and also the benefits of pursuing a beaten enemy. This was not generally done and the tactic was regarded as quite bloodthirsty and revolutionary at the time.

Frederick the Great tended to ignore cavalry in favour of infantry. This was reflected by the famous poor showing at the battle of Mollwitz (1741), where they suffered severely at the hands of the Austrians, the finest cavalry of the day. Gradually, however, Frederick worked on the cavalry and by 1756 had produced an excellent mounted arm with which to support his foot. The Prussian cavalry formed in three lines with cuirassiers (heavy cavalry) in the fore, supported by a second line of dragoons (medium cavalry) and finally hussars (light cavalry).

The British cavalry contingent at the Battle of Minden (1759) did not acquit themselves well under the command of the Marquis of Gransby, due to an apparent loss of nerve by the overall commander Lord Sackville. Instead, Prussian and Hanoverian cavalry stole all the glory which might have come the way of the British.

Britain's role on the mainland of Europe during the war was minimal so no real enduring tests were put on the cavalry. Britain's troops were employed in prising away French colonial possessions, and in battles such as Plassey (1757) cavalry played no part. In the action in Canada, culminating in the taking of Quebec, neither the British under General Wolfe, nor the French under General Montcalm had any cavalry with them, due to the difficult nature of the terrain.

Terrain also affected the use of cavalry in the American War of Independence. In all the campaigns the British Army used only the 15th and 16th Light Dragoons as mounted troops. A unit of Brunswick Dragoons arrived in 1776 as part of the Hessian mercenary contingent but without horses or saddlery, even though they were dressed as cavalry. Mounts were never forthcoming and the 'Dragoons' campaigned as infantry.

Impressed by the British Light Dragoons, the Americans raised four regiments for their own army. Dressed in varying coloured coats, the troopers performed well enough but were no match for their British counterparts.

Rules for 18th century wargaming

As Stuart has said in the previous chapter, the principal development affecting warfare in this period was the bayonet, making each musketeer his own pikeman and reducing the vulnerability of close-order infantry to cavalry. However, if cavalry could attack infantry from flank or rear they would still massacre them, so a formation known as the 'square' was developed from the 16th century schiltron. In this, the respective companies or divisions in a regiment moved to face outwards in a hollow square, with the officers and Colours in the centre. Artillery gunners menaced by cavalry would retire into the safety of the nearest square. Cavalry cannot mêlée with infantry in a formed square. It takes a wargames move to form square. Formed cavalry from now on may not charge down any slope greater than one in ten without becoming disordered.

Although flintlock muskets in this period showed a great improvement over the clumsy weapons of the 16th and 17th centuries, for our wargames purposes we still say that infantry may move *or* fire, but not both, within a single move. All infantry now count as MI apart from trained skirmishers—LI. Body armour has all but disappeared apart from breast and back plates on cavalry. There are no SHC; cuirassiers count as HC, dragoons, etc, as MC and hussars and light dragoons as LC.

Muskets have an improved range of up to 200 paces but deduct −1 from their dice score when firing at over 120 paces.

Artillery has become lighter and more manoeuvrable, and horse artillery, which can move at cavalry speeds, has been introduced. Light guns (3 to 6 pdrs) can now move at 120 paces a minute, but medium guns (6 to 9 pdrs) at 60 and heavy guns at 40, as before.

Ranges and fire factors remain unchanged except that guns can now fire canister shot at short range *if* the player wishes and remembers to write out the appropriate order. Add +2 when firing canister at a close-order target, +1 against an extended order unit (artillery gunners always count as being in extended order).

When mêléeing with infantry bayonets, count as 'other' weapon in the table on page 40.

Chapter 5

Napoleonic warfare
by Bruce Quarrie

The Napoleonic Wars exercise a peculiar fascination all of their own amongst a very high proportion of wargamers. In particular they were a period of evolution and reform in warfare. From their outset the proud armies of Austria, Prussia and Russia, which had been moulded under the harsh discipline and rigid tactical thinking of an altogether slower and in some ways more gentlemanly age, were often dismayed by the comparatively undisciplined ranks of the French Revolutionary forces. Out of this maelstrom emerged some of the finest military commanders of all time—Napoleon and his Marshals: the dashing Lannes and the avaricious Masséna; the flamboyant Murat and Ney; the dogged Davout; and many others. And in the devastating campaigns of 1805 against Austria and Russia, and 1806 against Prussia, the French swept all before them.

These defeats in turn brought out the talents of such men as Barclay de Tolly in Russia, the epileptic Archduke Charles in Austria, and Scharnhorst and Gneisenau in Prussia. Vigorous reforms were initiated in all their armies, but these were insufficient to bring victory in the various campaigns of 1807 and 1809—although the Austrians nearly succeeded at Aspern-Essling in May of the latter year.

Meanwhile, Napoleon had ill-advisedly sought to place his eldest brother on the throne of Spain. A British expeditionary force defeated the French at Vimiero in 1808 but was then forced to retire in the face of vastly superior numbers, ending in the retreat to and embarkation from Corunna, where Sir John Moore lost his life. However, the British were soon back again under Sir Arthur Wellesley but, after defeating the French at Oporto and Talavera, they were forced to retire to the security of their strong fortifications at Torres Vedras, in Portugal.

Up to this time, the British Army was the only one in Europe which had seriously succeeded in discommoding the otherwise victorious French. This was due to several factors, but the most important was that the British, having the recent experience of the American War of Independence to draw upon, had initiated a reform programme long before any of their Continental counterparts. This gave them greater mobility and firepower as well as the beginnings of the most effective light infantry force in the world.

The situation remained relatively static throughout 1810-11 in Europe but the French commanders in Spain, Masséna and Soult, suffered further rebuffs at the hands of the British and their increasingly competent Portuguese allies.

The Iberian Peninsula became, in fact, a real ulcer in the French flank, drawing ever greater numbers of troops into its inhospitable maw where they were plagued by disease, malnutrition and the decidedly uncivilised attentions of Spanish guerrilla bands.

Thus it was hardly under the most auspicious circumstances that Napoleon began planning his next move—an invasion of Russia itself. The Grande Armée assembled for this was the largest single force the world had ever seen, or would see again until the bloodbath of the First World War. Nearly half a million men, the majority of them coming from the host of minor European states which had fallen under French domination, marched across Germany and the infertile Polish plains and crossed the River Niemen in June 1812. The Russian forces fell back before them, although there were a large number of minor clashes and a more important one at Smolensk, before they finally made a stand outside Moscow at Borodino. After the bloodiest battle of his career, Napoleon finally forced the Russians to retire, but he had not succeeded in destroying them. The belated entry into Moscow followed as a rather hollow triumph.

Napoleonic warfare

Napoleon, with his forces sadly depleted, now had the choice of wintering in Moscow and continuing the campaign the following spring, or ordering a withdrawal to more hospitable quarters. In the end he adopted the latter course, but he had left it too late. Winter set in and the weary, undernourished and poorly clad soldiers of the Grande Armée began to die like flies from cold, starvation and attacks by roving bands of cossacks—Russian irregular light cavalry. The Russians opposed the retreat at every opportunity and by the time the French reached the River Beresina they were a sorry shadow of their former might. Whole units had been wiped out in their entirety and discipline had all but disappeared. Napoleon's Russian gamble had failed.

Heartened by these events, the Austrians, Russians and Prussians formed a new coalition in 1813 which was aimed at wiping out the French threat once and for all, and restoring the exiled Bourbon monarchy.

The resources available for Napoleon to draw on were meagre. He had for some time been drawing on 'Classes' of conscripts in advance, creating a 'manpower overdraft', and, like Nazi Germany in 1944-45, was now increasingly forced to employ young boys and older men, to the dissaffection of the home populace. Nevertheless, by the time the Allied coalition forces crossed the River Elbe, he was able to meet them with a new force of over 120,000 men.

Astonishingly, the superior Allied forces met defeat after defeat—at Lützen, Bautzen and Dresden in particular. By this time, however, it was not just the French army which was feeling the strain of nearly 20 years of almost continuous warfare. The commanders were also becoming tired, dispirited and, inevitably, careless. The drama drew to a climax at Leipzig, in what is now East Germany, in October 1813. The French and their allies occupied a large semi-circle east of the city, where they were assaulted by nearly twice their own numbers of Austrians, Prussians, Russians and Swedes under the overall command of the Austrian, Schwarzenberg.

A massive see-saw battle developed, both sides securing local advantages but

Left *Hinchliffe figure designer Peter Gilder advances his cavalry during a battle at Northern Militaire in 1976.* **Below** *A general view of the same engagement showing some of the beautifully sculpted terrain for which Peter is famous.*

unable to make a real breakthrough. The 'crunch' occurred when Napoleon's Saxon contingent deserted, leaving a gap in his line. This caused the French to shorten their front and begin retreating; but there was only one bridge over the river to the west of the town and this was blown up in the heat of the moment at just the wrong time, leaving large numbers of French troops to fall into Allied captivity or be drowned trying to swim, as was the Polish Marshal Poniatowski.

The end was now in sight but Napoleon refused to give up. Throughout the months following Leipzig, French troops continued to give battle and, indeed, to defeat their opponents time after time in localised encounters. But they were never able to follow up victories because of manpower shortages, especially amongst the cavalry, and the war reached its inevitable conclusion. Napoleon was forced to abdicate by his own Marshals and was exiled to the island of Elba, with a token honour guard.

Thus the wars should have ended, were it not for Napoleon's ambition, determination and sheer drive—for he was, by this time, a sick man. Instead, while the Bourbons ruled from Paris, the 'master' plotted his return. This occurred in 1815, when he returned to mainland France, gathered loyal allies around him, and marched on Paris. The Bourbons fled and once again Napoleon was ruler of France.

The Allies in their complacency had disbanded most of their armed forces; and, while Napoleon was recruiting vigorously, his enemies idled. Only the Prussians, who had perhaps suffered the worst ignominies under the French yoke, were truly awake, and even they suffered defeat at the battle of Ligny. The small British contingent still garrisoned in the Low Countries, however, managed to stall the French at Quatre Bras, and so laid the scene for the last act—Waterloo.

Here, the French displayed all their usual courage and, had not their leaders—in particular Ney and Napoleon himself—been so 'off form', could have carried the day. Instead, the cream of the French cavalry was expended in a fruitless exercise; Blücher's Prussians, recovered from Ligny, took them in flank; the Old Guard was forced back in disorder; and Napoleon lost his final battle. Exiled to St Helena, he died in 1821.

Why Napoleonic?

Exciting stuff, yes; but not yet sufficient to explain its appeal to 20th century wargamers. To get the easy part over with first, much of it has to do with uniforms. There have been many periods in history—particularly the 16th and 17th centuries described earlier by George Gush—when military costume has been so flamboyant and colourful as to almost defy description. And the Napoleonic period is certainly colourful. Moreover, as far as uniform references go, it is probably the best documented era in military history; which makes researching and painting your model soldiers very much easier. Bright colours predominated: yellow, red, blue and green, mostly in almost primary shades, outfitted the troops of the time. They wore shakos (hats) with elaborate badges, trimmings and plumes; jackets and waistcoats of various colours ornamented with bright metal buttons, elaborate epaulettes (shoulder straps), lanyards, gorgets and belts; steel breastplates and flamboyant fur-lined pelisses (short, embroidered, cold-weather coats); bearskins and brass, steel or leather helmets; knee-length breeches and long gaiters or boots. The whole concept of camouflage was virtually unknown. Uniforms were designed so as to make troops recognisable, both to their friends and their enemies. Napoleon suffered agonies at Waterloo because he could not make out whether the blue of the forces approaching from his right were those of French or Prussian infantry!

The budding wargamer interested in this period is, literally, faced with a bewildering array of choices. He is also—and this is one of the main attractions of the period—virtually guaranteed of a more-or-less local opponent against whom he can fight on a regular basis. If I was starting afresh, I know what I would go for; and I leave it to the reader to choose for himself: a 15 mm scale army of one of the minor states—Italy, Poland, Bavaria, Wurttemberg, Saxony, Westphalia or Portugal for preference.

As I said in the introduction, 15 mm is the coming scale, and is also relatively inexpensive compared with 25 mm, although you have to choose your manufacturer. Moreover, the vast majority of Napoleonic wargamers start with a French or a British army; others recognise this fact and, desirous of finding opponents, start creating Austrian, Prussian or Russian forces. The big advantage in choosing one of the smaller states, however; particularly in this scale; is that you can recreate the *entire* army at a relatively low cost and time investment.

However, uniforms and affiliated subjects are only one of the reasons why people become interested in the Napoleonic period. Another which I have already mentioned is the charisma of the central characters. Consider Napoleon himself, for example: born on the island of Corsica in 1769, he became a solitary but self-willed child with a devouring interest in history—particularly military history. Possessed of a mathematical flair, he was accordingly seconded to an artillery unit, where his latent ability to 'win friends and influence people' first became really apparent. To begin with, he was deeply immersed in the Corsican struggle for independence, but this was soon put behind him when he received promotion as a result of his success during the siege of Toulon. After a short spell in prison . . . one of the hazards of being successful during a period of civil turmoil . . . he restored himself to the existing government's good books by turning his guns on a rampaging mob and, shortly afterwards, was appointed commander of the rag-tag French 'Army of Italy'. Here, he turned potential disaster into triumph after narrowly defeating the Austrian invaders.

Appointed commander of the forces despatched to Egypt to annihilate the British presence in the Mediterranean, Napoleon fared no more successfully than did Rommel two centuries later, and returned to stage a coup d'état against the increasingly unpopular French government. From this he emerged as one of the three new rulers of France—called Consuls; a position which he rapidly consolidated, particularly following the lucky victory over the Austrians at Marengo. From First Consul it was a relatively easy step to get himself crowned as Emperor; which, combined with the assasination of the unfortunate Duc d'Enghien, was sufficient to rally the Austrians and Russians against France once more. The result was the French triumph over their combined forces at the classic battle of Austerlitz in December 1805—an encounter which every Napoleonic wargamer should certainly try; followed by the defeat of Prussia in 1806 after the double battle of Jena-Auerstadt, which left Napoleon virtual master of Europe—apart from the recalcitrant British.

Napoleon's domestic reforms—many of which survive to this day—are outside the scope of this volume; suffice to say that had he stopped here, in 1806, and contrived a peaceful solution to his European opposition, conditions

Above *Part of a Napoleonic wargame between Steve Tulk's Russians and Bruce Quarrie's French; here two French infantry battalions have been forced to form square by advancing cossacks, but lancers and horse grenadiers are coming to their rescue.* **Above right** *Russian infantry battalions advance in column of divisions and echelon formation against a French 12 pdr battery on a hill constructed from polystyrene ceiling tiles.*

in Europe today would unquestionably be very different. But his overwhelming ambition took control and led eventually, via umpteen thousand lives, to Waterloo.

On the face of it, Napoleon does not thus seem to be a particularly appealing character. Nevertheless, he had the necessary charisma of the born leader which induced men to follow him through extremes of physical and mental discomfort even whilst they cursed him.

He was no great military innovator. Most of what he accomplished was through tools which lay ready to hand when he assumed power. The linear formations of the Seven Years' War so ably described by Stuart Asquith in the preceding chapter relied principally on an ironclad discipline which left no room for individual initiative. The undisciplined rabble of the French revolutionary armies was largely incapable of fighting in this way. Moreover, there was not time to give them the extensive training in such things as musket drill which were so vital to that slow-moving, rather artificial, mode of warfare. As a result, they developed a new way of fighting in small, manoeuvrable columns of attack in which it was easier for untrained troops to maintain formation, screened by companies of light infantry whose job was to harrass the enemy and draw fire from the advancing columns behind.

Weapons and tactics

It must be remembered that, virtually throughout the Napoleonic Wars, the main tactical formation remained the line: battalions of infantry drawn up in three ranks deep (two ranks in the British Army), just as throughout the latter part of the previous century, whose principal weapon was close-range volley fire. In 1805-6 the French disrupted this cosy system by advancing in their

columns at great speed, preceded by light infantry in irregular formation who fired individually rather than in volleys. This upset the well-drilled regulars in the Austrian, Russian and Prussian ranks, causing them to open fire too early and to lose the well-oiled precision of their normal drill. When, however, the French encountered British troops who not only did not flinch at the skirmishers' harrassing tactics, but who sent forward their own light infantry to deal with them on an equal basis, they lost—because their columns could not equal the volley firepower of the stolid British line. Gradually the other European powers learned this lesson, and French victories became more and more marginal as *they* failed to adapt, leading to the eventual *débâcle* at Waterloo.

The musket used by virtually all infantrymen during the Napoleonic period was essentially unchanged from that used by Frederick the Great's troops and their opponents. A smoothbore muzzle-loader of no great range or accuracy, triggered by a flintlock system which became useless in wet weather and whose barrel became rapidly fouled by the coarse black powder used as a propellant charge, it was an effective weapon in the hands of well-trained troops. However, the loading and firing drill was complicated and easily forgotten in the heat of battle, resulting in many accidents and misfires; while the vast clouds of smoke it produced rapidly reduced visibility to little more than a few yards.

The average bayonet fitted to Napoleonic muskets was no better. Usually of only partially tempered steel, it looked formidable in bright daylight but was of more practical use in roasting barbecues than in skewering the enemy. As a consequence, battle records show comparatively few bayonet wounds, and in keeping with this a good set of Napoleonic wargames rules should discourage infantry mêlées.

Rifled muzzle-loaders such as the British Baker rifle were more effective, since the barrel grooves imparted a spin to the ball, enhancing both range and accuracy. However, they were slower to load than smoothbores and were only issued in relatively small numbers to specially trained formations. The same

Above *The cossacks have been pushed back and now it is the Russian infantry's turn to be intimidated as the French infantry and cavalry advance upon them.* **Below right** *The Russian infantry have formed square and the bulk of the French cavalry is hastening to the other flank where a dangerous situation has arisen. The wargames table is a sheet of chipboard laid on top of the dining room table.*

applied to the very effective, but expensive to manufacture, air rifles used by selected Austrian troops.

In the field of artillery, things had improved since the battles outlined in the previous chapter. Gun carriages had become lighter and hence more mobile, and an increasing proportion of batteries were designated as horse artillery, their crews being mounted. However, the mainstay was the foot artillery—heavy guns and howitzers of 8 pdr weight upward—which could only be manoeuvred with difficulty and which therefore tended to remain where they were positioned at the beginning of a battle.

The main type of ammunition fired by all guns of this period remained roundshot, or cannon balls. The balls used in field artillery ranged from 3 lb to 24 lb in weight, although heavier ammunition was used in siege guns. As in the muskets, the propellant was black powder, pre-packed in bags of the required size. This still gave a respectable velocity and range—up to around 1,800 yards in the case of a 12 pdr gun. The havoc caused by these balls on the densely packed formations of the day can be imagined, as they did not stop at first impact but continued bouncing and rolling in many cases for several hundred yards—and woe betide the ignorant soul who put his foot out to stop one as if it were a football!

The second most common form of ammunition was canister. This was a bagged round of musket balls which came out of the gun barrel at high speed and spread in a cone shape, like a shotgun cartridge. At close range—up to around 600 yards for a 12 pdr—canister was absolutely devastating.

The third type was shell, which was only fired from howitzers—short-barrelled cannon firing on a high trajectory. Shells were circular iron balls containing an explosive charge and a fuze which sometimes exploded over the desired target, but which was notoriously inaccurate except when used against very large formations where the gunners simply could not miss. A variation on shell, used solely by the British during this period, was shrapnel, which combined the virtues of shell and canister.

Gun barrels were made of brass or iron, mounted on sturdy wooden carriages held together with iron bands, and with iron-rimmed wooden wheels. They were towed by teams of (usually) six or eight horses, their trails being attached to a limber for this purpose. When in action, artillery pieces were generally deployed in batteries, normally of six or eight guns, although the Russians used much larger batteries. Depending on its size, each piece was crewed by upwards of five men, excluding the drivers.

Field artillery was sub-divided into two categories: foot and horse. Foot artillery batteries contained the heavier 8 pdrs and upwards, and howitzers. As the name implies, the crews marched on foot at the same pace as the regular infantry. In horse artillery batteries, all the crew was mounted, and due to the comparatively light weight of the pieces, these guns could keep up with cavalry.

The infantry in this period were essentially divided into line and light, although within these groupings there were several sub-divisions. Basically, the task of the line was to fight shoulder to shoulder in what is known as 'close order', while the light infantry operated in much looser formation as skirmishers. Light troops were usually picked from the best marksmen as well as the more intelligent and agile men. A further category is the grenadier and, although the use of the grenade had fallen out of favour except during sieges, these men retained an élite status as shock troops and were usually selected from the tallest and strongest men in a regiment.

An infantry regiment in the field, depending on nationality, would be composed of one or more battalions, while each battalion would comprise four or more companies, each averaging 140 men at full establishment. The norm was six companies to a battalion, of which one would be a light company and

one composed of grenadiers; and two to four battalions to a regiment. Frequently, however, the grenadiers and light infantry would be assembled into independent 'battalions' on the battlefield for particular tasks. You must appreciate, however, that these remarks are very general and that there were exceptions to every rule.

Cavalry were divided between heavy, medium and light. The 'heavies'—cuirassiers, horse grenadiers, carabiniers and heavy dragoons—were used as shock troops in boot-to-boot charges. They were the equivalent of the grenadiers in the infantry and were usually big men mounted on large horses, frequently wearing heavy brass helmets and steel breastplates and wielding heavy, more-or-less straight swords. The medium cavalry was essentially composed of dragoons, whose original function, you will remember, was as mounted infantry. This role had gradually been eroded by this time, and most dragoons fought as mounted cavalry, being able to fulfill the tasks of either heavy cavalry in charges or as light cavalry in reconnaissance and pursuit.

The light cavalry were the darlings of the army, comprising chasseurs, light dragoons and hussars, normally dressed in particularly flamboyant and colourful uniforms bedecked with lace and fur. Their essential role was in the pursuit of a defeated foe, as they could move faster than any other troop type, and for this reason they were usually held in reserve during a battle.

Falling midway between medium and light cavalry were the lancers or uhlans. They were usually converted dragoon regiments, a proportion of whose strength was equipped and trained in the use of the lance. As well as being particularly valuable in the pursuit, the reach of their weapons enabled them to counter-charge heavy cavalry with a good chance of success, although once enmeshed in a mêlée they had to revert to their light cavalry sabres. However, because of their versatility, they are among the most popular troops seen in Napoleonic wargames armies. Virtually all European armies used them with the exception of England.

Cavalry regiments were organised into squadrons which correspond to the infantry companies. A regiment would have four or more squadrons, each again of around 140 troopers.

Formations and strategy

The formations adopted by both infantry and cavalry depended on their function at any particular moment during a battle. On the approach, both infantry and cavalry would normally be in 'column of march', with the artillery and baggage train in the middle of the road, the infantry to either side, and the cavalry at the head and tail with flankers and scouts out on reconnaissance. Arriving on the battlefield, the infantry would deploy into more compact 'columns of companies' or 'columns of divisions' (a division being a two-company frontage). This gave them great manoeuvering flexibility. The cavalry would initially form up in columns of squadrons, only deploying into lines when they were about to go into a charge. The infantry could take one of two courses when approaching the enemy: form into line to use their muskets, or press on at full speed in column in an endeavour to use shock tactics to break through. Meanwhile, the foot artillery would usually be deployed pretty well to the rear to give covering fire and to weaken the enemy lines at chosen points, while the horse artillery would be well to the fore. Ideally, all three arms would advance in unison, the cavalry to handle the enemy's mounted troops and then force his infantry to form into squares*, which would then be hammered by the horse artillery until they were sufficiently reduced that one's own infantry could charge and break them. In practice, events rarely followed such a neat pattern! Theory is all very well, but what is the opposing commander doing all this while?

Battlefield strategy, better known as Grand Tactics, was in all essentials the same as throughout the 18th century. Following initial contact (unless your opponent was in a prepared position, in which case you had more time to plan), part of the army would be used to 'pin' the enemy, to try to hold him to his positions whilst friendly troops were deployed for a flank attack—still the most effective way of winning a battle with least casualties. As this manoeuvre became revealed, the enemy commander would be obliged to use up his reserves or swing part of his line to meet the threat. Concentrated artillery fire would then be poured upon those units forming the 'hinge', while the main body of the army, formed up in assault columns, would march inexorably to contact. Frequently their opponents, weakened and demoralised by the heavy fire they had sustained, would break and rout before physical contact was made. This would be the moment to send the cavalry in, and the opposing commander would have no option but to choose between flight and surrender.

That is the 'textbook' Napoleonic battle, but there were many variations upon it. Success also depended to a considerable extent in superiority of numbers, particularly in cavalry and artillery. To this end, Napoleon devised his remarkably flexible Corps system. A Corps was a miniature army, containing its own integral infantry, cavalry and artillery. The French army on campaign would consist of several such Corps, not necessarily all of the same strength, marching in a 'chequer board' formation with each Corps ideally no more than a day's march from two or three supporting Corps.

When one Corps made contact with the enemy army, it would immediately hurry forward to pin it down, while urgent messages were despatched to Napoleon himself (who would normally be travelling with the Imperial Guard, which itself grew eventually to Corps size), and to flanking Corps commanders. They would hasten towards the battlefield and, hopefully, one or more of them would be able to approach on the enemy's flank or rear.

If faced with two opponents, as he often was, Napoleon's usual concern was to prevent them uniting. To this end he would detail off one or two Corps to pin one of the opposing armies, then force-march the greater part of his army to engage the second at superior odds. Having dealt with this, he would then be able to march back to the relief of his other Corps, defeating the second enemy by weight of numbers. In war it is local superiority which counts rather than overall numbers, as the Germans were to demonstrate in 1940 and 1941.

For the wargamer, then, the Napoleonic era presents boundless opportunities for the display of different skills. The least satisfactory type of game is one in which both sides have absolutely equal forces and the terrain is devised to give equal opportunities to each player, as this type of battle gives the least opportunity for creative generalship—one reason I personally do not enjoy competition play. Refighting historical battles with the forces scaled down in

---

*The only way infantry could protect themselves effectively against charging cavalry in this period was to form their regiments into hollow square formations which, lacking flanks or vulnerable rear, could rarely be broken. Ney lost the cream of the French cavalry at Waterloo by futilely charging round and round the unweakened British squares, as he had no artillery or infantry support.

proportion to their real numbers is much more rewarding because you can analyse what you think the opposing commanders *should* have done and then try it out, often with surprising results. Waterloo, for example, will be won by the French nine times out of ten. Games in which the forces are unequal also have their attractions. One player could, for example, have an extra regiment of cavalry while his opponent is given an additional artillery battery. Success will then depend upon which player makes best use of his own superiority. Similarly, unbalanced games in which one player is given a small force a third or half the size of his opponent's, and has to fight a 'rearguard' action for a given length of time, can be most stimulating. It is all up to your own creativity.

Rules for Napoleonic wargaming

Since the Napoleonic period is essentially an extension of the 18th century, the foregoing rules for that period are really all that one needs. The only special troop types which need to be mentioned are the lancer or uhlan, a light cavalryman armed with a long lance, and the rifleman. Lancers move as LC but fight as 'cavalry with long spear' in the first round of a mêlée, as cavalry with 'other' weapon in successive rounds of mêlée; the lance in this period is essentially a 'one shot' weapon.

Lancers were known on occasion to break infantry squares, although it was still a very rare occurrence, and we must give them the opportunity in our rules. Unlike other types of cavalry, they *may* attempt to mêlée with an infantry square so long as their morale is 8 or above and the infantry's is below 8 *at the time of impact.* You thus test the morale of both units just before the cavalry make contact, and after they have probably received casualties from musket fire. If they fail, the lancers wheel and flow around the sides of the square. You can thus see that the odds of such a charge succeeding are fairly remote, as it was historically.

The best way to use cavalry against infantry in this period is to bring them up within sight of the infantry, forcing them into square; then use horse artillery to weaken the square, and finally bring up your own infantry to charge home with the bayonet while your opponent can only bring a quarter of his men to bear on yours.

The rifled musket was the other significant weapon which saw widespread introduction during the Napoleonic Wars. Although used during the American War of Independence, the rifle was essentially a new 'invention' during this period. It was muzzle-loaded and thus slower to fire than smoothbore muskets, but had a longer range and greater accuracy. Riflemen count 0–100 as short range and add +1 to their fire factors, 100–200 as medium range with no additions or deductions, and 200–300 as long range with a −1 deduction. Riflemen move as LI.

Chapter 6

The Crimean War
by Don Fosten

After the abdication of Napoleon Bonaparte the four major Allied Powers, England, Austria, Prussia and Russia, formed the Quadruple Alliance which sought to see Europe settle down to a long period of peace and economic stability.

For the wargamer it seems at first glance to be a barren period and yet it was a time when most of the great powers continued to be engaged in either bloody internal strife or civil wars and, in the British case, in an almost continual Colonial campaign.

Between 1830-48, the French invaded North Africa and conquered Algeria and, as a consequence, formed 'Turco' native regiments and raised the Foreign Legion. Between 1818 and 1848, British contingents were engaged in fighting either Kaffirs or Dutch and German settlers in South Africa. In Spain the Carlist War of 1834-39 saw the Foreign Legion in action together with a British Legion of some 9,600 officers and men. In Italy revolutionary ferment between 1820 and 1847 culminated in their Great War of Independence between 1848-9.

There was an insurrection in Portugal between 1823-4, which was followed by the savage Miguelite War of 1828-34, in which Clinton commanded a further British contingent of 5,000 officers and men. In 1848, there was a bloody rising in Hungary which led to the raising of the exotically uniformed 'Honved', the national army, and the Austrian Army, considered by many historians to be the best European army of the period, was involved not only with the Hungarian uprising, but internal trouble in her Italian possessions, and with Prussia. Prussia suffered a revolt in Berlin which was followed by trouble in Schleswig-Holstein and this led to the Frankfurt Convention of German States.

Even little Belgium suffered an equally fierce revolution in 1830, and Greece fought her long War of Independence between 1821-1832, during which many British officers were once again involved. The Turks had not only to sustain an almost continuous campaign with Russia but also a war with Egypt in 1832.

As I have said, for Britain the period was one of almost continual strife somewhere in the world. Conflicts ranged from the Mahratta War in 1817-18 and the Burma War of 1823-26, to the conquest of Scinde, the 1st and 2nd Sikh Wars and the 1st Afghan War. From 1839-42, Britain was also involved with the Chinese and French in the 1st Opium War.

Even the United States of America did not escape conflict. She was fighting the Seminole Indians in the swamps of Florida, the Plains Indians, the Texas

War of Independence from 1835-36 and the Mexican War of 1846-48, including incidents such as the defence of the Alamo.

Britain's first major involvement since Waterloo came in 1854. The Russians had used religion as an excuse and had occupied the principalities of Moldavia and Wallachia and the Allies decided to support the Ottoman Empire. Allied fleets passed through the Dardanelles and Bosphorus and drove the Russian fleet into Sevastopol harbour.

In the Crimean War which followed it was the Allied-Turkish strategy to overcome the Sevastopol garrison and neutralise the fleet while holding the Russian field army at bay.

The Russians were commanded by General Prince Alexander Menschikoff, the French by Marshal Armand de St Anaud, later by General Canrobert, and the British by Field Marshal Lord Raglan.

The great battles of the Crimean War—the Alma, Balaclava, Inkerman and the great siege—cost thousands of lives, yet the outcome was never decisive. At the Alma the British lost 2,000 killed all ranks, at Inkerman 597 killed, including three Generals and 39 officers all told. In that fight the Russians lost 12,000 including 256 officers. At Balaclava General Scarlett's Heavy Brigade drubbed 12,000 Russian cavalry in what a French officer claimed was the finest charge he had ever seen, but Lord Raglan charged and lost the Light Brigade in a disaster which has gone down in history as a glorious mistake. Some 675 officers and men rode to almost certain death without question and gained nothing. They had nearly 430 dead and almost all their magnificent horses were either killed or had to be destroyed, yet this never to be forgotten event took only 20 minutes from the charge to the retreat.

During this short but terrible war the British lost nearly 20,000 officers and men but ironically 16,000 of them died from disease or from maltreatment in their own hospitals. However, a great deal had been learnt. Florence Nightingale and her staff began to open new vistas of military hygiene and medical care and the campaign resulted in major advances in logistic technology and uniform and equipment reforms.

Until the Crimean War little change had occurred. In 1816, the British had adopted the lance, a concession to the prowess of the Polish and French lancers, the Prussian uhlans, and Bosniaken and the Russian cossacks.

In 1841, the Prussians had experimented with a breech-loading rifle and during the same period the old Land service muskets of the British had been adapted to percussion caps. These were shortly followed by the short-lived Brunswick rifle, then by the more efficient Minié and, in May 1854, by the excellent Enfield.

Artillery had changed little and the British Horse and Field batteries of the Crimea differed little from either the draught harness and tackle or the limbers and carriages of those used at Waterloo.

Immediately after Waterloo most European nations adopted variations in each other's uniforms. The British 'Waterloo' shako disappeared, to be replaced by the almost universal pattern bell-topped shako. Collars became higher and tighter, waists higher, coat tails longer and epaulettes and wings more and more exaggerated. Hussars and cuirassiers were dressed in very expensive uniforms and equipment, resplendent with gold and silver lace, tight breeches and huge muff caps and crested helmets. Much economy had taken place before the Crimean War began but even so the British still clung to the coatee, the shako and the old crossbelt equipment with knapsacks supported by constricting shoulder belts and breast harness. Conversely the Russians had simplified much of their uniforms and in the field their men wore thick ankle-length grey overcoats and ugly ball-topped helmets or shapeless undress caps. By the time the war ended the British had introduced tunics, simpler headdress and modified equipment.

Rules for mid-19th century wargaming

Playing rules as for the Napoleonic era with the addition of breech-loading rifles and guns. Add +1 to previous fire factors at all ranges for these weapons except as noted below. Ranges were, in fact, substantially greater in many cases, but *effective* ranges (ie, those at which weapons would normally be fired) virtually unchanged. A higher proportion of LI should be allowed for in drawing up army lists. The infantry in this period were armed with a mixture of old muzzle-loading muskets and rifles, and the new breech-loaders in the proportion of approximately 6, 3 and 1 in 10 respectively, and this should be allowed for in drawing up the forces for a battle. Muskets and muzzle-loading rifles fire as previously.

As far as artillery is concerned, treat smoothbore muzzle loaders as in the 18th century but increase the ranges of rifled guns by a third and give them the above-mentioned fire factor at all ranges except when firing canister.

Chapter 7

The American Civil and Franco-Prussian Wars
by Terence Wise

The Industrial Revolution, which began in Britain in the late 18th century, inevitably had a dramatic effect on methods of waging war, and during the 19th century unprecedented changes in weaponry and radical new thinking on strategy, brought about by the new technological developments, completely revolutionised the art of war. The Industrial Revolution reached France *circa* 1830, Germany after 1850, and the United States of America around 1860. In the American Civil War and Franco-Prussian War, therefore, we have the first modern wars, the first total wars in the current sense, with national economies fully integrated into war effort. Furthermore, the change from essentially agricultural economies to industrialisation released a far greater percentage of a nation's manpower than had been possible in the past, and enormous armies could now be raised—armed, equipped, clothed and supported in the field—by the development of national war industries. For the first time the French Revolutionary concept of the 'nation in arms' was eclipsed by a new concept—the nation at war.

The American Civil War

In addition to being the first truly modern war, the ACW is usually credited with a long list of other firsts: the first appearance on the battlefield of large quantities of rifled small-arms and artillery, and of repeating rifles; the first use of the new steam engine, in the form of locomotives on the railroads, as a major means of transporting men and supplies; the first widespread use of the new telegraphic communications; the first extensive use of trench warfare; and the first use of machine-guns.

Despite these revolutions in weaponry, transportation and communications, at the outbreak of war in 1861 military thinking in North America was still dominated by the example of Napoleon, and consequently the strategy, tactics and even weapons initially employed by both the Confederate (CSA) and Federal (USA) armies were basically those of the Napoleonic era. The men marched off to war in uniforms as colourful and fancy as any of the Napoleonic Wars, armed with smoothbore, muzzle-loading muskets and artillery. They formed up in tight ranks and closed with the enemy for a musketry duel at short range, followed by a swift bayonet charge if successful.

This was the situation in the first half of the war, but gradually the introduction and perfection of new weapons brought about a change to more stagnant warfare, with one side taking up a defensive position and the other side unable to capture it, even with odds of three to one, without suffering casualties so severe as to make any victory Pyrrhic. The battle of Gettysburg (July 1–3 1863) is often quoted as the turning point.

Organisation

The official strength of the infantry regiments raised by both sides for the Civil War was between 869 and 1,049 men, divided into ten companies and with 35 officers. However, in practice regiments often had only 700 to 800 men, and after their first battle were frequently reduced to an average of around 500. The CSA fed new recruits into existing regiments, keeping them at a reasonable strength, but the USA allowed regiments to fall as low as 150 to 200 men, when they were disbanded and the survivors distributed amongst new regiments. This meant that on average the CSA regiments tended to be stronger than most Union ones and were usually more effective as a fighting force. One exception to this system in the Union Army was the Wisconsin regiments, where recruits were fed into existing regiments as in the Confederate Army.

At the start of the war the Union regular army contained some 16,000 infantry in ten regiments of two battalions each. In May 1861 the US Government raised nine more regular regiments, also to be of two or more battalions, each battalion having eight companies. These were the 11th to 19th Regiments. Full strength was never achieved and one battalion of eight companies was the norm for these regiments.

A few regiments were raised which were of a slightly different nature to the regular and volunteer regiments described above. For example, both sides had Zouave regiments, which were usually armed with the M1862 Remington muzzle-loading rifle with a faster rate of fire and longer effective range than the ordinary musket. In 1861 the Union also raised the 1st and 2nd Sharpshooters, each of four companies. These men fought mainly in small groups, sniping at officers and gun crews, and both regiments served with the Army of the Potomac. They were combined to form a single regiment in 1864. The CSA had no sharpshooter units as such but did employ a few snipers; perhaps no more than one hundred at any one place or time.

The regiments were grouped into brigades of from three to six regiments. Two to five brigades might be assembled to form a division, and a corps comprised from two to four divisions. An army could consist of any number of corps. The brigades and divisions of the Confederate Army tended to have more regiments than the corresponding USA formations.

Cavalry regiments in the US Army consisted of six squadrons, each of two companies. At first each company had 100 men and officers, but in 1863 the regulation strength was changed to between 82 and 100 men and officers. As in the infantry, actual strengths tended to be lower, between 70 and 80 men of all ranks. In 1863 the squadron was abandoned and replaced by a cavalry battalion of four companies. There were six regiments from the pre-war regular army, all other regiments being state or volunteer ones.

In the first two years of the war the cavalry was sometimes attached to infantry divisions, but early in 1863 the Army of the Potomac reorganised its cavalry as a separate arm right up to corps level, and other armies subsequently followed suit. A brigade comprised four to six regiments, a division two to three brigades, and a corps two or three divisions.

The Confederate cavalry regiment had ten squadrons, each with a theoretical strength of 60 to 80 men and four officers. In practice a figure of around 50 men of all ranks per squadron was more common. Brigades were formed from between two and six regiments, divisions from two to six brigades.

Two other aspects of the cavalry which should be taken into account are skill and supply of remounts. In the first half of the war the Confederacy had born cavalrymen and a good supply of horses: the Union had neither. In the second half of the war the Union had both, but the Confederacy was short on horses and many a veteran cavalryman was forced to turn infantryman. To allow for this in wargaming it is best to give the Union Army a lower proportion of cavalry until 1863, the CSA a lower proportion in 1864–65. The proportion of cavalry to infantry should never exceed 1:6 for either side.

Union artillery batteries normally had six guns, but four or eight were used occasionally. The 12 pdr batteries usually had four 12 pdrs and two 24 pdr howitzers: 6 pdr batteries usually had four 6 pdrs and two 12 pdr howitzers. Later in the war batteries usually had all one type and calibre pieces. There were about ten crew members per gun, plus spare crews, drivers, etc.

The artillery was normally allotted at the rate of four batteries per infantry division, with often half this force withdrawn to form a corps reserve. Each army also had a reserve of light (horse) and heavy batteries. From Gettysburg on, however, the Army of the Potomac usually concentrated its artillery at corps level, with nine batteries per corps.

Confederate batteries were most often of four guns, sometimes six and occasionally eight. Crews were approximately as in the Union Army. Four CSA batteries made a battalion, and there was normally a battalion to each infantry division, with further battalions as corps reserve.

The proportion of guns to men should be no more than one gun for every 1,500 men in the Union forces; one gun for every 1,750 men in the CSA forces.

Each gun had a limber and caisson with ammunition, and each battery had reserve caissons, a baggage wagon per gun, and a field forge. This clutter of vehicles often caused problems on the battlefield, where the gun and caisson horse teams were drawn up behind the guns in action to a depth of 47 yards. These vehicles should therefore be represented on the wargames table by at least two-horse model limbers, and if possible four-horse caissons as well.

Weapons

At the outbreak of war the infantry on both sides were armed mainly with muzzle-loading, smoothbore muskets, but by the autumn of 1862 they were equipped with muzzle-loading rifles. Production of breech-loading rifles for the US Army was begun that same year, but their distribution was limited in 1863, and at The Wilderness (May 1864) Grant still had only three per cent of his army armed with the Spencer repeating rifle. Even in 1865 the number of regiments equipped with the repeating rifle was low, possibly no more than ten per cent.

In 1861 the Confederate infantry were armed with perhaps 85 per cent smoothbore and 15 per cent rifled muskets, compared with 75 and 25 per cent respectively for Union infantry. In 1862 there was approximate equality with 50 per cent single-shot rifles, and in 1863 with 100 per cent single shot rifles. However, the quality of the CSA infantry weapons was drastically inferior to those of the Union in 1864–65, as the US Army began receiving the new magazine rifle.

The American Civil and Franco-Prussian Wars

Infantry weapons

Weapon	Rounds per minute	Range in yards Max	Range in yards Effective	Range in yards Battle	Remarks
US Percussion M1842 muzzle-loading smoothbore	2-3	300	150	75	Standard weapon in 1st months of the war.
Springfield M1861 muzzle-loading rifle	3	1,000	500	250	Most common weapon on both sides.
Enfield muzzle-loading rifle	2-3	1,100	500	300	2nd most common, more accurate than Springfield.
Remington (Zouave) M1862 muzzle-loading rifle	3	1,200	600	350	3rd most popular, more accurate than Springfield.
Sharps breech-loading rifle M1863	8	1,800	600	350	Issued to US sharpshooters 1863.
Spencer breech-loading rifle	16	1,800	600	350	Union troops only. Those captured by CSA had limited use as CSA had no means of manufacturing the ammunition.
Henry breech-loading rifle	20	1,800	600	350	Faster than Spencer but more likely to jam.

Cavalry weapons

Weapon	Rounds per minute	Ranges in yards Effective	Ranges in yards Battle	Remarks
Enfield muzzle-loading rifled carbine	2-3	500	300	Widely used by both sides early in war. Preferred by CSA even late in war because of accuracy and rugged reliability.
Spencer breech-loading rifled carbine	10	450	300	Best cavalry arm of the war. Decisive in many cavalry actions.
Sharps M1859 & M1863 breech-loading rifled carbine	8	500	300	2nd only to Spencer.
Burnside breech-loading rifled carbine	about 8	450	250	3rd most popular.

At the beginning of the war the cavalry was armed with sabre and revolver, and at least two squadrons (companies) in each regiment also had muzzle-loading carbines or single-shot rifles. By early 1863 all cavalry was equipped with carbines and from 1864 there was a gradual switch to breech-loading repeaters in the Union Army. Confederate cavalry relied on the Enfield carbine with its greater range and accuracy.

Sabre fighting was mostly abandoned after the first two years of the war, although the US cavalry continued to carry sabres. Most cavalry now carried two revolvers for close-quarter fighting. These were mostly Colts, Remingtons and Starrs, all of which held five rounds (six if placing one under the hammer) and had a rate of fire of 15 rounds a minute. Reloading in the saddle during a mêlée was impracticable and, particularly in the Confederate cavalry, troopers often carried as many as four revolvers, and sometimes a shotgun. The maximum range of the revolvers was 300 yards, but effective range was only 50 yards, with an accurate range of 25 yards.

The main field artillery pieces used by the US Army were the three-inch Rodman and 20 pdr Parrott rifles, and 12 pdr Napoleon smoothbores. The Confederates used 12 pdr Napoleon smoothbores and 12 pdr Whitworth rifles in roughly equal numbers, and the 20 pdr Parrott rifle. By 1863 both sides usually had between a third and a half of their pieces in three-inch or 20 pdr rifles, the remainder smoothbores. The heavy batteries of both sides were armed with the 30 pdr Parrott rifle.

The new rifled guns could out-range the old smoothbore of the Napoleonic era, and could hit harder and with more accuracy, but there were 'teething' problems (malfunctioning of ammunition or fouling of the rifling) which caused loss of accuracy. When firing shell, rifles also tended to drive the projectiles so deep into the ground that their burst was ineffective. These faults to a large extent neutralised the technical superiority of rifled guns, although they retained the advantage of longer range.

Smoothbores were much less accurate and had a shorter range, but were rugged and reliable and, loaded with canister, were murderous weapons at close range. Both rifles and smoothbores could fire solid shot, shell and canister, although with varying degrees of success. Shot and canister were highly effective when fired by smoothbores, but shell was mainly fired by rifled guns and by howitzers. Solid shot was ineffective against the entrenched infantry of the second half of the war if fired from the range necessary to protect the gunners from the infantry's rifles, and therefore shell came to be used more and more in 1863-65. It was also employed as long-range canister against infantry attacks, and statistics show that in fact shell was no more effective than shot *unless* used for such plunging fire, hence the predominance of shell ammunition for howitzers.

Machine-guns were available to both sides during the war but were seldom used due to problems of mobility and ammunition supply. They played only a small part in the war and were limited mainly to sieges, in a static role where ammunition could usually be stockpiled in advance. The table shows known details of the various machine-guns' performances.

Tactics

The overall effect of introducing rifled guns and small-arms, breech-loading instead of muzzle-loading, and repeating rifles in place of single-shot, was to increase the volume of fire, range and accuracy of both the infantry's and artillery's firepower. In the first two years of the war the infantry and cavalry of both sides advanced into hails of iron, steel and lead in the old-fashioned formations of the Napoleonic era, and died by their thousands. Eventually the

Maximum effective ranges*

Artillery piece	Shot	Shell Min	Shell Max	Canister	Remarks
6 pdr Napoleon smooth-bore	1,000	250	800	250	Used mainly by CSA in 1861. Replaced as soon as possible by 12 pdr or 3-inch rifle.
12 pdr Napoleon smooth-bore	1,200	250	900	300	Most popular smooth-bore. Reliable and effective.
12 pdr Whitworth rifle	2,000	—	—	300	Used by CSA. Exceptionally accurate.
3" Rodman rifle	1,800	250	1,400	300	From 1863. Popular with Union Army, favourite of their light batteries.
20 pdr Parrott rifle	1,900	250	1,400	300	The basic piece, used by both sides.
30 pdr Parrott rifle	2,200	250	1,600	300	Used by heavy (reserve) batteries.
12 pdr smooth-bore howitzer	1,070	150	750	250	
24 pdr howitzer	1,300	300	800	300	
32 pdr howitzer	1,500	500	900	300	

*All at approximately 5 degree elevation.

Machine-guns

Weapon	Ammo	Rounds per min	Range	First used in action	Remarks
Billinghurst Requa Battery gun	.58 cal	7 volleys of 25 shots	Over 1,000 yds	Built late 1861	Loose powder used and hazards of sparks & rain.
Agar machine-gun	.58 cal	120	Unknown. Deadly at 800 yds	29/3/1862	US weapon; 2 taken by CSA June 1862. Unpopular because of unreliability.
Williams machine-gun	1 pound shell or canister	20	Up to 2,000 yds	31/5/1862	CSA weapon. 42 guns in 7 batteries. Very reliable but ammo problems restricted use.
Vandenberg volley gun	Various musket cal	—	90% hits at 100 yds	—	
Gatling machine-gun	.58 cal	250	—		12 purchased by General Butler in 1864; Hancock also 12, Porter 1. Not very accurate. Improved M1865 too late for Civil War.

Above left *A wargames reconstruction of the American Civil War battle of Murfreesboro; Confederate cavalry have cut the railroad to the rear of the Federal forces.* **Left** *The early stages of a game with Confederates on the left and their Federal opponents to the right. The river lengths, walls, houses and bridges are all vacuum-formed accessories from Micro-Mold.* **Below left** *Another triumph for the Confederates, this time at the battle of Bitter Creek.* **Above** *American Civil War artillery in action; the triangular bases give each gun's angle of fire. Cotton wool is being used to represent smoke. Note the limbers behind the guns.*

men refused to advance to certain death in such a manner, and forced on their commanders completely new tactics of dispersal and manoeuvre. By late 1863 neither infantry nor cavalry could or would launch a frontal assault against infantry and artillery in a good defensive position.

The new formations were spread out and flexible. Infantry regiments now fought in two-deep lines, with skirmishers forward to shield the main force and weaken that point of the enemy line to be attacked. Sometimes up to half a regiment might be used as skirmishers and, if a divisional attack was being launched, whole regiments might be committed in this role.

Thus attacks were launched in waves of regiments in double ranks, preceded by the skirmishers and with 250 to 300 yards between waves. Individual regiments often had two companies up to 500 yards in advance as skirmishers, six companies in the main line, and two companies in reserve 300 yards to the rear. Such spacing prevented heavy casualties from artillery and rifle fire, and permitted greater manoeuvreability than the old-fashioned columns.

If infantry came under heavy fire they usually went to ground and immediately dug in, or continued their advance in short rushes from cover to cover in small groups of men. In the second half of the war the foxholes and rifle pits were expanded into trench systems, bases from which the armies could manoeuvre, and most of Lee's victories were the result of his outstanding ability to use hastily thrown up field fortifications as a base for the aggressive employment of fire and movement.

The US Drill Manual for the ACW period states that infantry columns could move at 70 yards per minute in common time, 85 yards quick time, and 110 yards double quick time. On long marches the regiments marched in columns of four abreast, or sometimes by column of companies.

The cavalry of both armies manoeuvred in column of fours for maximum flexibility. Until 1862 their battlefield formation was two-deep, but in that year both sides changed to single line. With frontal charges against infantry and artillery unthinkable, mounted cavalry operations became limited almost entirely to screening and reconnaissance missions. On the battlefield cavalry were no longer used as a separate arm in support of infantry attacks, but were relegated to the role of mounted infantry. In this they used their mobility to seize advanced positions until the arrival of the infantry, or to cover gaps in the battle line, or to cover retreats. They normally operated as dismounted skirmishers, with a skirmish line ahead of the main body, horse holders to the rear, and occasionally a mounted company on each flank.

Cavalry carbines did not perform well when firing continuously for long periods, nor did they have the range of infantry weapons. This meant that cavalry could not put up a prolonged resistance to infantry attack, even after 1864 when the US cavalry began to be armed with repeating carbines.

Cavalry-versus-cavalry actions were still fought, usually raids or minor combats away from the main forces, but from 1863 carbines and revolvers took the place of sabres even in this fighting. Cavalry mêlées therefore tended to be rapid and confusing encounters, each side trying to hit hard and fast, and break away quickly. At the greatest cavalry battle of the war (Brandy Station, June 1863) 20,000 cavalry were engaged for more than 12 hours, and at the height of the battle charges and counter-charges were made continuously for almost three hours.

Once the bulk of the infantry was armed with rifles, the artillery was forced into a purely defensive role, for it could no longer advance in close support of infantry attacks, the enemy's infantry being able to pick off the gunners before they could get their guns within canister range. Therefore the artillery was normally used defensively, to break up infantry assaults before they moved into small-arms' range. The rifled batteries would open fire with shot at their longest range, next the smoothbores would join in with shot, then both would switch to shell, and finally to canister at close range. Canister was so effective that smoothbore pieces, which could fire canister twice as fast as rifled pieces, remained in service in both armies throughout the war and the 12 pdr Napoleon was the commonest gun of the war.

Counter-battery fire could be employed but the artillery was much better employed against infantry targets, and at Gettysburg the Union batteries were ordered to cease counter-battery fire so as to conserve ammunition and allow the guns to cool for the infantry attack which would follow the artillery duel.

The railroad network was used by both sides for the movement of supplies and troops to the front, or from front to front. Alongside this new system of transportation was the older, waterborne one, and rivers and river craft continued to play a vital role throughout the American Civil War.

Another new development which affected both strategy and tactics was the electric telegraph and cable which, while ensuring rapid communication between field commanders and government, also permitted politicians and political motives to hinder field operations. The almost instantaneous transmission of a war correspondent's dispatches, appearing in the daily press within hours, also proved a two-edged sword, for breaches of security frequently enabled one side to learn of their opponent's plans before those plans could be put into operation or carried to fruition.

At a tactical level the field telegraph was used to link divisions, corps and armies, and was in common use in the US Army and to a lesser extent the CSA Army long before the end of the war. The telegraph was also used to link aerial observers in captive balloons with the ground in order to pass information on the enemy's dispositions. Semaphore flags and the heliograph were still used to maintain communications between field HQs and the lower commands, and with reconnaissance forces and isolated detachments.

The Franco-Prussian War

In 1866 Prussia defeated the mighty Austro-Hungarian Empire in a whirlwind campaign known as the Seven Weeks' War. As a result of this victory Prussia was able to annex some neighbouring German states and force others to join the so-called North German Confederation. Four years later, in the summer of 1870, Prussia and her German allies totally destroyed the military power of the second French Empire in a single month of fierce and costly fighting, radically altering the European balance of power and making possible the unification of Germany under Prussian leadership.

France had led Europe in the military field since Napoleon's time, whereas Prussia had gone into a decline since 1815 and, until a decade before, had been the least of the Continent's major military powers. The completeness of the Prussian victory astounded the world. How had it come about? The incompetence of the French high command explains a great deal, but France's defeat was not due entirely to a faulty command system—her whole military system was at fault, geared to fight wars of the type waged in the first half of the 19th century.

In the five short years between the end of the ACW and outbreak of war in 1870, the solid shot and Minié ball had become obsolete, and the weapons which had fired them were museum pieces: the Minié ball had given way to the elongated, conical bullet of the rifle, and the artillery now fired long, streamlined shells with explosive charges, detonated by sophisticated fuzes. The Franco-Prussian War was to be the first war fought primarily with the breech-loading rifle, utilising one-piece cartridges containing projectile, propellant and primer, discharged by a firing pin, and in which the new steel, breech-loading rifled field pieces made their first appearance. But the lessons of the ACW had not been assimilated in Europe, and the French and Prussian armies were to learn the hard way that significant changes in tactics were necessary to counter the vastly increased firepower of infantry and artillery.

Organisation

The establishment strength of a Prussian infantry regiment was three battalions, each of 1,000 men and officers. Mobilisation of the Prussian Army in 1870 was carried out efficiently, and within 18 days 1,183,000 regulars and reservists were passed through barracks in Germany and embodied in the wartime army; 462,000 of them being transported by rail to the French frontier to open the campaign. Regiments in the front line were therefore at full strength initially.

The traditional divisions of grenadiers, musketeers and fusiliers within a battalion now meant very little in practice. Guard regiments naturally provided the élite of the army, while light infantry were supplied by the special schützen (marksmen) and jäger (hunter) battalions. The Landwehr (militia) was commanded by regular officers and NCOs and included veterans of the Seven Weeks' War. The military law of 1861 had made it compulsory for all able-bodied men to serve three years in the regular army, followed by four years in the reserve, with a short period of training annually thereafter, and for these reasons the Landwehr of 1870 was a far more effective force than the Landwehr of 1813-14. However, it was generally used for second line duties, although some regiments did see front line service in the quieter areas of the front.

Two regiments were organised as a brigade, and generally speaking two brigades formed a division. A corps normally consisted of two divisions plus a battalion of jägers, a pioneer battalion, an artillery regiment, two light cavalry regiments and a battalion of Train. The jägers were usually attached to one division, but each infantry division had its own regiment of dragoons or hussars, a pioneer company and a field 'division' of the artillery regiment, namely two 4 pdr and two 6 pdr foot batteries. Each corps was therefore a self-contained unit of around 30,000 men, with the following units under direct corps HQ command: the third 'division' of the foot artillery regiment, a 'division' of horse artillery (four 4 pdr horse batteries), the remainder of the pioneer battalion with their equipment Train, and the Train battalion with the ammunition and supply columns, ambulances, etc. The Landwehr was usually organised in separate divisions with auxiliary services and was not mixed with the regular troops.

The establishment strength of a cavalry regiment was 600 men and officers; regiments of heavy cavalry *(reiter)* were organised in divisions of three regiments. Artillery batteries normally had six pieces.

The official strength of a French Line infantry regiment was 2,400 men and officers, divided into three battalions, each of four or six companies. However, mobilisation of the French Army during 1870 was conducted in such an archaic way, and resulted in so much confusion on the railway network, that after three weeks only about half the reservists had reached their regiments and many of those lacked essential items of uniform and equipment. Regiments were therefore frequently as low as 1,500 men, half the size of the German equivalent.

Light infantry were represented by 20 chasseur battalions (each with six companies of 115), three tirailleur regiments (each of four battalions, with each battalion having six companies of 110), and three Zouave regiments (each of three battalions, with six companies of 95 per battalion).

The Imperial Guard formed an entire corps as in Napoleonic times, with three regiments of grenadiers and four of voltigeurs (each of three battalions, with seven companies of 95 per battalion), one Guard Chasseur battalion of ten companies of 79, and one Guard Zouave regiment of two battalions, each battalion having six companies of 95 men.

The Line infantry was also organised in corps, being formed from between two and four infantry divisions, each of two brigades, with two regiments per brigade. Each division had attached an engineer company, two 4 pdr and one *mitrailleuse* (machine-gun) batteries, and a battalion of chasseurs. Each corps had attached a cavalry division of three regiments, engineer and Train companies, and several artillery batteries of 4 pdrs and 12 pdrs. There was also a Reserve Cavalry Corps of three divisions, and a general Army Reserve of artillery and engineers. However, the confusion during mobilisation meant that several corps had no engineers or artillery, while others lost whole infantry regiments, which finished up in different corps!

The Guard cavalry regiments, and Line hussars, spahis, chasseurs d'Afrique and chasseur à cheval regiments, all had five squadrons, officially of 150 men and four officers, but often as few as 105. Line heavy and medium cavalry regiments had four squadrons. Artillery batteries normally had six pieces. The *mitrailleuse* batteries were classed as field artillery and had from one to six guns.

Following the final defeat of the French Army at Sedan, the Third Republic was formed and managed to raise 14 new army corps from surviving regiments, reservists, depot troops and volunteers, but these corps were usually understrength and even more subject to missing units—particularly the vital Trains. The irregular cavalry units of this period were usually of squadron strength.

Weapons

The Prussian army had adopted the single-shot Dreyse needle gun—so-called because in the firing action a needle penetrated the base of the cartridge—as early as 1848. This was a great advance in infantry weapons in its time, firing a breech-loaded, non-metallic but single-piece cartridge that was loaded by a bolt action. Although used in the short war with Denmark in 1864, it was not until the Seven Weeks' War that the rest of Europe awoke to the enormous advantages of such rifles which, because of their loading action, provided an infantry weapon which for the first time enabled all infantry to reload whilst *lying down*. Although a single-shot weapon, the Dreyse needle gun enabled the Prussian infantry to fire six rounds to every one from a muzzle-loader. However, by 1870 the Dreyse needle gun, despite modifications, was of an outdated design. Its two major faults were that the firing needle tended to break and the breech was not really gas-tight, which greatly reduced its range: maximum effective range was 600 yards, accurate range about half that. Each soldier carried 80 rounds. Dragoon and hussar regiments were armed with the Dreyse carbine.

After the 1866 *débâcle,* when the Austrians were almost literally swept from the field by the Prussian rifle fire, the other European powers hastened to equip their armies with breech-loaders. The French version, the single-shot Chassepot rifle, was put into production in 1866 on the specific orders of Napoleon III, and by 1870 a million were available for the army. The Chassepot solved the problem of a gas-tight breech by introducing a rubber ring in the breech. This made the rifle safer and easier to fire than Dreyse's rifle, substantially increased the range, enabled a smaller calibre bullet to be used (increasing a man's ready ammunition to 90 rounds), and gave better accuracy. The Chassepot was sighted up to 1,600 yards and accurate to 1,000 yards at five to six rounds per minute. Dragoons, hussars, chasseurs à cheval and chasseurs d'Afrique were armed with the Chassepot carbine.

The French Army was the first to be totally equipped with rifled cannon, as early as 1858. These were the traditional bronze cannon, muzzle-loaders like their predecessors, but rifled. Rate of fire was three to four rounds a minute, but the shells were fitted with the old-fashioned time fuzes, set to explode only at 1,200 and 2,800 yards and subject to all the erratic results experienced in Napoleonic shells. Like Napoleonic artillery pieces, these guns were also heavy and difficult to manoeuvre.

The 4 pdr field piece fired an 8 lb shell a maximum of 3,000 yards but effective range was only 1,200 yards, and the 12 pdr a 24 lb shell to 3,700 yards maximum, with an actual effective range of 2,800 yards.

In the Seven Weeks' War the Prussian artillery was equipped partly with smoothbores and partly with the new rifles. By 1870, however, it had been completely re-armed, and all field batteries now had the new cast steel breech-loading field pieces from the Krupps works. (Steel guns had not been adopted by other major powers at this time because the casting process tended to produce flaws in the barrels, and the pieces were therefore considered unreliable.) The steel 4 pdr cannon had a maximum range of 3,300 yards, firing an 8.5 lb shell, the 6 pdr a maximum range of 4,000 yards with a 14.7 lb shell. Effective ranges were only slightly less and rate of fire was the same as the French pieces. In addition to the much longer effective range, the German artillery also had the advantage of using percussion fuzes, which exploded shells on impact and eradicated the inefficiencies of the old time fuze.

The French had one secret weapon—the *mitrailleuse,* an early form of machine-gun which consisted of 25 barrels in a group, each detonated by turning a handle. The mitrailleuse had a range of nearly 2,000 yards and a rate of fire of 150 rounds, or six discharges, per minute. Mounted on a carriage for mobility, it was about the same size as the 4 pdr cannon.

Tactics

The battalion column was still the basic deployment formation in both armies at the outbreak of war, although it had generally been recognised that the new weapons would transform battlefield tactics and that the skirmish line would have to be strengthened at the expense of the columns held back for the final assault. In 1866 the Prussians had demonstrated the devastating advantages of rifle-armed infantry in a well-sited defensive position, but they had found that in battle the assault columns tended to disappear into the skirmish line, and after the Seven Weeks' War they therefore restored close order formations. The French general staff concluded from the lessons of 1866 that they must find and hold good defensive positions, and in their attacks bypass any enemy defensive positions. Their chasseur battalions, however, were trained in speed, agility and initiative, to encourage flexible formations on the battlefield which could make full use of all available cover.

However, in the main the infantry of both sides attacked in waves on battalion front, with each company in two lines, preceded by a skirmish line. Casualties were appalling—the Prussian Guards, for example, losing 8,000 men in 20 minutes when making a frontal assault on the cemetery of St-Privat, on the French right flank at the battle of Gravelotte.

A different picture began to emerge after the last of the great battles (Sedan), by which time the commanders had learned the hard way. A perfect example of the new tactics for infantry occurred on October 30, during the siege of Paris, when Prussian Guards attacked the village of Le Bourget. Although advancing across open fields and under Chassepot fire from the village and shell-fire from supporting forts, the Guards succeeded in outflanking both sides of the position and taking 1,200 prisoners for a loss of 500 dead. In this combat *all* the Prussian companies were dispersed into a loose and widely spaced skirmish line, and advanced by bounds, supporting each other by fire, making the best possible use of cover and offering only small targets. It was a textbook example of the infantry tactics of the 20th century.

Light cavalry was used only in the reconnaissance and screening roles, but heavy cavalry continued to be employed in the shock role of the Napoleonic Wars—the charge *en masse.* Such charges were made in two-deep squadron lines, but they could now be received by infantry in line, and several French cuirassier regiments were virtually wiped out when charging unbroken Prussian infantry at the battle of Worth. Even the most successful charge of the war, Von Bredow's Death Ride at the battle of Vionville, achieved its objective only at the cost of 380 men out of 800.

Prussian artillery was handled extremely boldly and effectively throughout the war, with the divisional batteries placed right in the front line with the infantry to break up attacks and support counter-attacks. It was also switched to crucial points quickly during battles, providing local superiority when and where needed. This bold handling, together with the greater range and accuracy of its guns, frequently gave the German artillery a decisive role in battles.

By comparison the French artillery was rather feebly handled, with divisional artillery frequently far in the rear and rarely co-ordinated to provide concentrated fire. The French command also retained a general artillery reserve, which was rarely used to support anticipated break-throughs, but rather deprived the infantry and divisional artillery of much needed support, allowing the divisional artillery to be destroyed piecemeal by the German counter-battery fire. This reserve also tended to clog up the vital lateral roads to the rear during a battle.

Even the *mitrailleuse* failed to give the results anticipated, mainly because the weapon was too often treated as an artillery piece, sited as a battery and used at long range. This enabled the Germans, who had a healthy respect for the weapon when encountered in small groups in concealed positions, to locate and destroy many *mitrailleuse* batteries before they could become effective.

Railways and telegraphs had been in widespread use before the war and both sides used them to facilitate the movement and control of their large armies. However, both sides still relied to a great extent on the traditional foot-slogging for movement between battles once concentration in a region had been achieved, and the limited role of railways *at the front* at this date is aptly illustrated by the fact that only the capture of French supplies saved the invading Prussian armies from a condition bordering on starvation.

Rules for late 19th century wargaming

As previously, except that Confederates and Prussians should have a higher proportion of LI than their Federal or French opponents, and CSA cavalry count as irregulars. Rules could be introduced for machine-guns if you wish, using the table on page 103 but, as Terry has said, they only played a small part in the ACW so can be ignored without sacrificing realism; the *mitrailleuse* should be treated as a rifled artillery piece firing canister at all ranges.

Chapter 8

Colonial warfare
by Ted Herbert

Colonial wargaming is a fascinating branch of miniature warfare that is steadily expanding in popularity. At wargaming exhibitions, one often sees a gallant band of red-coated figures in dire straits as they fight off one attack after another from zealous Zulus or fanatical followers of the Mahdi. But what exactly do we mean by 'colonial actions' or 'small wars' as they used to be called in Victorian and Edwardian times? A 1906 account* defined small wars as 'all campaigns other than those where both the opposing armies consist of regular troops . . . conditions are so diversified, the enemy's mode of fighting is often so peculiar, the theatres of operations present such singular features, that irregular warfare must be carried out in a method totally different from the stereotyped system'. Under this definition we have a tremendous variety of actions and troops.

The campaigns

The major campaigns that spring to mind are the Zulu War of 1879—the colonial wargamer's scenario *par excellence;* the Second Afghan War of 1878-80 and the numerous skirmishes on the North-West Frontier of British India; the Egyptian War of 1882, the Sudanese War of 1884-85 and the Reconquest of the Sudan from 1896-99; and the two Boer Wars of 1880-81 and 1899-1902. These are the 'big four' periods in colonial wargaming, not the least contributory factor to their popularity being the ready availability of several excellent ranges of wargames figures for these periods. However, the colonial era as a whole encompasses far more than these famous British campaigns. In the scramble for colonial possessions at the end of the 19th century, such units as the French Foreign Legion, the Italian *Bersaglieri,* the German *Schutztruppe,* the Russian Cossacks, the American 7th Cavalry, the Portuguese Colonial Regiment, the Dutch Marines and the Belgian *Force Publique* were all engaged in actions against irregular opponents. The action with the most variety of troop types was the Boxer Rising of 1900 when troops of eight nationalities defended the Legation Quarter in Peking.

Going back to earlier times, we can bring in Clive's campaign in 1757 against the Nabob of Bengal; the French Indian Wars in North America from 1754-59; Wellesley's campaigns against the Mahrattas in 1798-1803; the Sikh Wars of

---

Small Wars—their Principles and Practice, Captain C.E. Caldwell, HMSO, London 1906, reprinted by EP Publishing Limited in 1976.

Colonial warfare

1845–46 and 1848–49; and, of course, the Indian Mutiny of 1857–59. Some of these actions lie on the margins of our definition of small wars, since both sides contained regular troops, to a greater or lesser extent, but they are definitely colonial in flavour.

One attraction of the colonial period is the opportunity to paint not only British redcoats but also many other colourful colonial uniforms. The Indian Army *circa* 1900, for example, was one of the best dressed armies of all time. It is true that by this date khaki campaign uniforms were in general use but then how many Napoleonic wargamers paint their troops in campaign rather than full dress? British troops can be painted in red tunics up to at least 1885, when it is believed that they were worn for the last time in action during the Battle of Ginnis in the Sudan. Some of the troops landing in South Africa in 1899 wore the home service uniform of red coats and blue trousers but it is very doubtful if they actually fought in it, despite the apparent evidence of contemporary prints, which were designed to inspire patriotic feeling at home. As regards standards, the last occasion on which they are known to have been carried into action was at Laing's Nek in the First Boer War in 1881; a year later an order was published forbidding the taking of Colours on active service.

There were many causes—real or supposed—of small wars. Some were straightforward campaigns of conquest or annexation, for example the Russian invasion of territory beyond the Caspian in the Central Asian desert; others were police actions to suppress or punish insurrection, for example the Morant Bay rebellion in Jamaica in 1865; campaigns to wipe out an insult or avenge a wrong, for example, the Abyssinian campaign in 1867–68 to rescue European prisoners held under duress by King Theodore; wars of 'self defence' against a powerful neighbour, as in the Zulu War of 1879, when the British authorities held that the Zulus were threatening Natal; expeditions in areas such as the North-West Frontier to follow up marauders, protect a loyal village, or raid an outlaw stronghold; or campaigns of political expediency, for example, the Gordon Relief Expedition that Gladstone was obliged by public opinion to mount in 1885.

These varying causes of conflict did not, however, extend to direct clashes between European troops. Partition of territory by mutual agreement was the

The Zulu War, 1879. The 24th Foot fend off an attack

normal practice. Fictitious conflicts between European forces can, of course, be devised for the wargames table; and this helps considerably in the development of a balanced colonial game. Two of the most plausible actions are a clash between Anglo/Egyptian and French forces at Fashoda in the Sudan in 1898 or an invasion across the passes of the North-West Frontier by the Russians, with British and Indian units being hastily mobilised to defend the Raj. One actual dispute over territorial rights was the battle between British and Portuguese forces in 1891 at Chua Hill, on the border between Rhodesia and Portuguese East Africa. This was the only conflict between Great Britain and another European State between the Crimean War and the outbreak of World War 1.

Although the European troops almost always possessed better arms and better discipline than their irregular opponents, they suffered severe casualties once the assegais of the Zulus, the hatchets of the Maoris or the swords, spears and knives of the Mahdists (Dervishes) came to close terms. Moreover, the over-stretched European troops had the disadvantage of fighting far away from home; often they were badly led and supplied and had poor knowledge of the local terrain. They did not overcome the tribesmen easily and sometimes they did not overcome them at all. At Isandhlwana in Zululand, six full companies of the 24th Foot (2nd Warwickshire) were annihilated; and at Maiwand in Afghanistan the 66th Foot (Royal Berkshire) suffered very heavy casualties; whilst in 1896 at Adowa in Abyssinia over 6,000 Italian troops were killed or wounded, together with 4,000 of their native allies.

Tremendous difficulties over commissariat and transport faced the Victorian generals. In the Abyssinian campaign, Sir Robert Napier had to transport his supplies, including 500,000 lb of biscuits, 100,000 lb of salted meat and 30,000 gallons of rum, across a fierce desert and along a route extending for more than 400 miles through roadless mountain terrain; 10,000 fighting men required 26,000 baggage and riding animals and were accompanied by 12,000 followers. This campaign, by the way, saw the ordnance of Europe transported across the mountains of Africa by the elephants of Asia. When the Russians advanced on Khiva in Central Asia in 1874, their force of 5,500 men required 8,800 camels. Even Sir Frederick Roberts' column that achieved the famous march of over 300 miles from Kabul to Kandahar in only 20 days, with supplies reduced to an absolute minimum, needed nearly 8,600 transport animals to support 10,000 men. The guarding of lines of communication was particularly expensive in manpower; in 1880 some 15,000 men maintained the Khyber line between Peshawar and Kabul, leaving only 12,000 men for the Kabul Field Force.

The solution eventually found to this problem—and one amenable to use on the wargames table—was the establishment of small flying columns, as used by the French in Algeria, the US Army in the Indian Wars and the British in 1897 against the Mahdists at Abu Hamed. A typical column in hill warfare consisted of four or five battalions of infantry, a troop of cavalry, a mountain battery, a field company of engineers and supplies for four to five days.

Terrain

One factor that it is vital to consider in colonial wargames is the nature of the terrain. The difficulties of inadequate knowledge of local terrain are illustrated by the Dutch campaign against the Sultan of Achin in Sumatra; the Dutch suffered a reverse because they failed to realise that they had reached the Sultan's stronghold at Kota Raja and withdrew when an immediate attack

Colonial warfare

The North-West Frontier, 1878. An Afghan gun battery holding the pass at Charasiah is attacked in the flank by Ghurkas.

would probably have been successful. Depending on the scenario chosen for a wargame, the terrain to be represented can vary from the coastal bush-covered plains and broken hill country of Zululand to the Ashanti bush consisting of dense undergrowth surmounted by thick stands of trees. Tribesmen were extremely well adapted to their natural environment, be it harsh desert or steaming bush, but they suffered as much as the Europeans in unfavourable conditions; once the Abyssinians, for example, descended from their highlands they moved very slowly in the hot and malarious country of the southern Sudan.

Some tips on the construction of terrain are given in the introduction to this book and these can be adapted to cover colonial games. A sand-coloured or grass-green cloth placed over boxes or books is quite effective; and pieces of cork, bark, lichen and stones can be strewn about the table to provide plenty of cover and give opportunities for hidden movement on the part of the tribesmen. A particular feature of encampments in the Sudan was a protecting *zariba* of thorn bush branches; whilst on the North-West Frontier circular breastworks of stone or timber, called *sangars,* were used as observation posts and defensive cover. Rocky slopes can also be constructed in irregular contoured layers from sheet cork, chipboard, insulation board or 'strawboard' painted in appropriate colours*. Sand, grit or pieces of bark can be glued to the layers to give a more effective finish. Expanded polystyrene is a rather fragile material to use, particularly if you intend to transport your terrain to wargaming exhibitions. It does, however, have the advantage of being very light in weight.

*See, for example, 'Those North-West Frontier Rocky Slopes!' by George Erik in *Wargamer's Newsletter* for June 1978, no 195, page 14.

The problems of the period . . .

It is apparent from what has been said so far that we are faced with a considerable problem in devising a set of wargaming rules to cover the huge range of actions, troop types and terrain falling under the heading of colonial warfare. I think it is fair to say that this is probably the main reason why colonial wargaming has not yet 'taken off' to the extent that other more manageable periods have. Unlike, say, Napoleonic wargames, where almost any set of rules will give a balanced game, since the opposing forces are based on approximately the same organisation and weaponry, colonial rules need much more thought and play-testing in order to provide balanced games.

Nevertheless, this is not a feature unique to colonial wargaming. Ancient wargames also have a large disparity in the weapons of opposing troops and in their discipline and motivation; yet a nationally accepted set of rules is available for this period. It should, therefore, be possible to develop a similar set for colonial games. However, at present most colonial players tend to favour the writing of rules to cover specific actions such as the Zulu War. The eventual solution may well lie in the development of an overall framework, with individual modules added to this for particular battles or campaigns. As far as the National Wargames Convention are concerned, Trevor Halsall's 'Rules for the Colonial Wars' have been used in the years in which an individual colonial championship has been held (1972 and 1973 at Leicester, 1974 at Nottingham and 1978 at Reading). It has been found that the best approach is for each player in the competition to play two games against each opponent, one with a European force and one with an irregular force; the points are then totalled to give as fair a result as possible. Under these rules irregular armies such as a Boer force composed entirely of marksmen or a Mahdist force of fanatics have proved very successful.

Because of the superiority in weaponry of the European forces, colonial wargames have to be set up rather differently from conventional wargames, in order to avoid a one-sided contest. This can be done, as in the Ancient period, by using a points system whereby a shieldless and unhappy tribesman armed only with a sword and a prayer is given a much lower points value than a regular infantryman armed with a magazine rifle. Trevor Halsall's Rules include such a points system.

Alternatively, clearly defined objectives that take account of the relative strengths and weaknesses of the two sides can be agreed on at the start of the game. The objectives do not have to be positive ones; they could be to prevent the enemy from achieving something within a certain number of moves or of not losing more than a certain percentage of casualties throughout the game. For example, the *doyen* of colonial wargamers, Douglas Johnson, suggests* that a European force could be given the task of starting at one end of the board and getting off the opposite end within a reasonable move limit and within an assigned casualty ceiling. The European units are, therefore, required to take the initiative and cannot afford the time to take up tactically advantageous positions such as static squares or entrenched lines. In such a game, the tribesmen may also be subject to a casualty ceiling and to avoid heavy losses may feint, skirmish and generally try to hold up the opposing column on the

---

*'Wargames of Survival' by Douglas H. Johnson in *Savage and Soldier,* Volume X, no 4, October–December 1978, page 9.

Colonial warfare

march, rather than attack directly. Once the time limit is up and the European force has not achieved its objective or else has suffered casualties beyond the agreed limit, the tribesmen win the game, provided of course that their casualty restrictions have been met. Another possibility is for the tribesmen to try to get off the board with as few casualties as possible, or perhaps with a herd of cattle or a wagon-load of supplies intact. They must try to retreat while the European troops seek to force a pitched battle. Obviously in this type of game careful thought needs to be put into the selection of terrain and the positioning of troops. It is a truism to say that the more effort you put into the preparation of a game the more enjoyment and satisfaction you get out of it.

In a campaign covering a wide area of country the advantages of the tribesmen in outscouting the enemy and achieving an element of surprise become even more pronounced. We have only to look at the build-up to Isandhlwana to see how *strategy*, with plenty of room to manoeuvre, favoured the Zulus. On this occasion it cannot even be said that *tactics* favoured the British. The map for a campaign might consist of varied terrain, with perhaps a series of villages or water-holes in tribal territory. The objective of the European force would be to bring the tribesmen to battle and to reduce their overall strength by a minimum percentage over the campaign as a whole. The Europeans would gain little from the occupation of a village or water-hole, although they might burn the former or poison the latter in order to claim points for the morale effect of a successful punitive expedition. On the other hand, the tribesmen would lose face (and points) for each village or water-hole occupied by the enemy. The total number of terrain points, including perhaps points for holding the heights or important observation posts, might equal the casualty points the Europeans need to inflict on the tribesmen to win.

In the course of the campaign, each side would have to decide how to divide their forces in order to garrison places under their control to prevent them being captured or recaptured by the enemy, while at the same time leaving sufficient strength to provide mobile striking forces that can reinforce key sectors. Limited observation and slower mobility for the Europeans, combined with hidden movement for the tribesman (using markers, some of which represent real units and some of which are dummies), can provide exciting and equally matched campaigns. In addition, attrition rates can be introduced to allow for the effect of sickness on the European troops. 'General Fever' was one of the main allies of the tribesmen.

... and ways of overcoming them

We have now reached a position where we can consider some of the background information necessary for the drawing up of detailed rules for the type of action that you may wish to recreate on the wargames table. In view of the difficulties already explained—and in any case in my experience many wargamers tend to be somewhat opinionated individuals who prefer to develop their own system—it is not my purpose here to suggest how a model set of rules should be formulated. What I can do, however, is to provide a few pointers that may be of use in developing your own rules or in modifying existing ones.

One feature in writing colonial rules is that we can carry over elements from other periods. The overlap of some aspects with ancient warfare has already been mentioned. In fact I once used the Ancient Wargames Rules of the Wargames Research Group to refight the Battle of Kandahar, a decisive action in the closing stages of the Afghan War from 1878–80*. Most of the Afghans were classed as D type barbarians to represent their volatile nature, Indians as Class C regulars and the British as Class B regulars. Effective rifle range was set at 24 in and artillery range at 48 in on a playing board measuring 8 ft × 6 ft. The rules seemed to survive the transition in period quite well; this is perhaps not so surprising when it is recognised that certain factors, such as infantry move distance, basic mêlée weapons and the distinction between trained and untrained troops, have not changed a great deal over the ages, as our editor mentions earlier. The rules reproduced very well the aggressiveness of the Afghans, as long as the prospects were good, and their gradual deterioration in morale when subjected to steady pressure from the British force. At the end of the game casualties on both sides amounted to about 25 per cent of those engaged but the Afghans were in no fit state to continue the fight and it was clearly a decisive victory for the British. The casualty rate was unrealistically high, at least on the British side, but in a campaign most of the casualties could be counted as lightly wounded men who would recover in time to take part in the next action.

There are also parallels with the Napoleonic period. Face-to-face with tribesmen of the fighting calibre of the Zulus and the Mahdists, the regulars were forced to re-adopt old orders of battle rendered obsolete in European warfare. If one wanted to be unkind, one could say indeed that some British generals had never given them up. Since the tactics of the tribesmen were essentially aggressive, the regulars were forced to conform to the enemy's method of attack. In view of the propensity of many irregulars for such unfair tactics as sweeping round the flanks of their opponents, the regulars formed a Napoleonic square as a defensive measure, even to the extent of developing an 'elastic' square for moving through dense bush. In addition, the old system of volley firing proved an effective way of halting determined enemy rushes; and cavalry armed with lances came back into fashion as an effective means of pursuing a routed enemy. On occasion, for example at Elandslaagte, lancers proved effective even against the Boers (who regarded the use of cold steel as unsporting).

Because of the relatively low firepower of the tribesmen, British battalions were able to adapt the drill-book attack formation of the time, which allowed for a skirmishing line of two companies with the remaining six companies acting as supports and reserves, to put as many as six companies into the firing line. In this revised system, the skirmishing companies were extended in sections, with one section from each in support 25–75 yards behind the firing line, while the two reserve companies remained in close order 100–200 yards behind the supports. When closing to mêlée, the reserves would move to the flanks and the whole line would close up for a Napoleonic-type charge.

Weapons

Turning now to weaponry, we can divide the colonial period into three main parts delineated by the availability of a) smoothbore and rifled muskets, b) single-shot breech-loading rifles and c) magazine rifles and machine-guns. The rifled musket was paramount in the British Army until about 1868, when one of

---

*'The Battle of Kandahar' by Ted Herbert in *Wargamer's Newsletter* for November 1970, no 104, page 3.

Colonial warfare

the first breech-loaders, the Snider-Enfield, was issued for general use; this was superseded in 1871 by the Martini-Henry, which continued in use until 1888 when it was replaced by the Lee-Metford magazine rifle; this in turn gave way to the Lee-Enfield rifle from about 1895 onwards. The Indian Army, in common with other colonial forces, lagged behind in these changes and did not receive the Snider-Enfield until 1871, the Martini-Henry until 1891 and the Lee-Metford until 1901.

With the introduction of magazine rifles and Maxim guns, the character of colonial warfare changed markedly. Before 1890 there were numerous instances of tribesmen charging regular troops sword in hand and in broad daylight (night actions were rarer as the sun never set on the British Empire!). However, after this date there was a growing disinclination on the part of the tribesmen to charge into close contact and there was more emphasis on long-range sniper fire. Even when tribesmen did charge into mêlée, the real shock effect came from a heavy burst of covering fire just before the attack went in. This change in tactics should come about naturally if appropriate weapon factors are used.

In assessing the firing factor for, say, a Martini-Henry rifle, account should be taken of the very high expenditure of ammunition that was required to score a hit. The supposed maximum rate of fire from an expert shot with this rifle was 23 rounds per minute but even with hasty aiming this rate would be at least halved; under battle conditions, owing to the overheating of the barrel and the need to conserve ammunition, the rate of fire would drop to a few shots a minute. At the major battles of Ulundi in the Zulu War and Ahmed Khel in the Second Afghan War, each man on the British side fired on average only ten rounds altogether; and during the Russians' disastrous attack on the Turkoman fortress of Geok Tepe, when General Lomakin's infantry were engaged for several hours, each man fired under 100 rounds. Statistically, it required about 35-40 rounds to kill or incapacitate an enemy: at Rorke's Drift more than 20,000 rounds were fired in an exceptionally long engagement but only about 500 Zulus were killed. So your fire effect should not be too devastating; a casualty rate in a close order charge to contact of up to two tribesmen for each infantryman firing is probably about right. With magazine rifles the fire effect should be increased at short range and in a two-hour battle the rate of expenditure of ammunition might go up from 50 rounds or so to 80 rounds.

Incidentally, while we are on the subject of statistics it is important to stress the advisability of consulting as many sources as possible before coming to any conclusions on which to base your wargaming rules. Even in such a well known action as Isandhlwana there is a common misconception that the battle was lost due to an administrative error, namely (to quote one recent source) 'the British Infantry ran out of ammunition because some idiot had forgotten to supply the tools for breaking open the reserve ammunition boxes'. In fact Dave Langley* has recently calculated that, based on average rates of fire, the Line infantry would still each have had about 17 rounds left, discounting the supply of any reserve ammunition, when the Zulus broke through. More importantly, the ammunition boxes were of Service Pattern Mark V or VI, the central portion of the lid of which could be opened by a single blow from a rock or a boot on the edge of the lid farthest away from the screw. The lack of screwdrivers was not

---

*'Isandhlwana: a Theoretical Examination' by Dave Langley in *Savage and Soldier*, Volume X, no 1, page 10.

therefore a serious handicap. The superiority of the Zulus in numbers was such that it is doubtful if the British could have won with an unlimited supply of ammunition, given the length of the initial perimeter line. So that is just a mild note of warning about not taking everything you read at face value.

The introduction of magazine rifles enabled regular troops to stand in line, rather than square, to meet the onrush of hostile tribesmen; and the Italians were very proud of their victory at Agordat in 1893, when they stood in line to repel Dervish attacks. In view of the superiority of European weaponry after 1890, the period at the turn of the century is not an easy time to write playable wargame rules for, although the Boer War of 1899-1902 provides an exception and there were occasional European defeats such as at Adowa in 1896.

Turning now to artillery, we can divide the Victorian period into four main stages of development comprising a) the smoothbore muzzle-loading guns and howitzers produced before 1860, with a maximum range of about 1,000 yards and a firing rate of two rounds a minute; b) the rifled breech-loading guns of 1860-77, which were not too dissimilar in effect from the rifled muzzle-loading guns (a retrograde step!) of 1877-1885; c) the advanced breech-loaders of 1885-1900; and d) the powerful quick-firing guns thereafter, with a range of 8,000 yards and a rate of fire of up to 50 rounds per minute. Wargame rules that aim to be comprehensive need to cover horse, field, heavy and mountain artillery for each of these main stages of development. In addition, they should cover two types of machine-gun: the early crank-operated forms such as the Gatling, *mitrailleuse* and Gardner; and the later trigger-operated forms such as the Maxim.

However, you may prefer to concentrate on one of the typically colonial guns, for example the 2.5-in rifled muzzle-loading screw gun immortalised by Kipling. This fired a 7 lb shell and was introduced in 1878. It was carried on mule-back in five loads, the breech and chase being screwed together on assembly. A typical ammunition mix for action on the North-West Frontier was 32 double common shell, 180 shrapnel, 55 case shot and 60 rounds of star shell. A high proportion of rounds fired were shrapnel since this type of fire was greatly feared by the tribesmen and provided useful support for a piquet or rearguard caught on the wrong foot by a charge of tribesmen. Mountain guns were very effective with plunging fire downhill but had a more limited effect firing uphill on *sangars*. The burst circle for a shrapnel shell was about 20-30 yards in diameter.

Setting up a wargame

Having looked very briefly at characteristic weaponry of the colonial period, let us now consider how we can actually set up wargame units and see how we can give them orders and assess how these orders will be carried out. There are four main figure scales available—15 mm, 25 mm, 30 mm and 54 mm. The latter two sizes of figure are used mostly for skirmish games, which we will consider later. Most of the mass action games that you will see at wargaming exhibitions will be in 25 mm scale, although 15 mm is catching on fast. Let us therefore assume for the moment that you wish to set up a colonial force using 25 mm figures. An excellent ground scale at this level is 1 mm represents 1 yard, with a man-to-figure ratio of between 20-1 and 33-1. A figure with a frontage of 12-15 mm can then represent an infantry section drawn up in two lines, each man having a frontage of about a yard (in parade formation it would be less). With four

Colonial warfare

The Sudan, 1885. The Black Watch in a zariba *by the Nile come under attack from Mahdist spearmen,* jihadiyya *and camelry.*

sections to a company and eight companies to a battalion, a full-strength battalion therefore consists of 32 figures, plus a commanding officer if desired. Actual numbers may, of course, be less than this due to casualties or sickness. A battalion may operate in two wings of up to 16 figures to give more flexible units. For Russian, Italian and French battalions, there should be two figures to a section, with four sections to a company and four companies in a battalion; whilst Belgian and German units, owing to a different organisation, should have two figures to a section and three sections to a company, but again four companies in a battalion. These suggestions are based on the organisations described in the respective drill-books of the time.

A 25 mm cavalry figure can represent two lines of horsemen forming a half-troop, a regiment consisting of four squadrons, each squadron comprising two troops. At full strength, a cavalry regiment is therefore composed of 16 figures, plus a commanding officer if desired. As regards tactics, the 1876 Cavalry Regulations for the British Army placed emphasis on flexibility of manoeuvre and on the importance of dismounted carbine fire. Where a classic cavalry charge was ordered, this took place in three successive waves. The first wave attempted to draw enemy fire; the second wave, 150–200 yards behind, manoeuvred to attack from the flank; and the third wave acted as a reserve and supported the first wave in charging home on weak points in the enemy's front. That at least was the theory.

Under the same system, one artillery figure can represent the crew of one gun plus drivers, with six figures and two gun models to a battery. In addition, a field battery should have five horse figures, a horse battery six and a mountain battery four ordnance mules and two baggage mules. A field company of engineers can be represented by a mounted officer and five dismounted sappers. A mounted infantry company can consist of four mounted men, of whom one holds the horses while the others dismount to fire. To represent this you can either have replacement dismounted figures or else glue 'skis' on the riders so that they stand up when they are detached from their horses. Special formations such as the Camel Corps can be scaled down as appropriate.

Above *Skirmish game set in Mexico, 1863. French Foreign legionnaires defend a ruined building.* **Above right** *The Sahara. A mobile square of legionnaires and zouaves marches to the relief of a desert fort, harried by Arab camelry.*

Regular infantry can be mounted on bases singly or in twos, in order to represent skirmishing ability, but if you prefer stands of three or four figures you should also have some single figures to act as skirmishers and to allow casualties to be deducted, as in other wargaming periods. Cavalry are best mounted in ones or twos in order to represent the formations of the time and also to enable them to travel in a narrow column of route. A suitable frontage is 20 mm for a cavalry figure and 25 mm for a camel.

Tribesmen have no particular formations, apart from the highly disciplined Zulus, and in general should have wider frontages than the regulars in order to represent their lack of cohesion. They can be massed in tribal groups under coloured banners. Sometimes at wargaming exhibitions one sees tribesmen advancing in immaculate Napoleonic lines and columns. This is most unrealistic; their formations should be broken up into dispersed groups. British cavalry, in fact, often found it difficult to charge straggling crowds of tribesmen as these could be ridden through without much effect; the tribesmen would simply melt away and flow round the flanks of the cavalry. One way of breaking up natives is to mount them on circular bases of varying size, the numbers counted in mêlée being calculated on a frontage corresponding to the diameter of the circles. This is not to say that some units, for example the Mahdist *jihadiyya* (non-Arab Sudanese trained originally by the Egyptians), were not capable of fighting in regular order but in general the mass of the tribesmen would be in very loose order. Provided that the bases are suitably decorated with fine sand and bits of bark to represent rocks, a large conglomeration of scattered warriors can look a fine sight. Zulus can, of course, have a frontage equal to that of the Europeans and fight in regular formations.

Having assembled our troops (and laboriously painted them) how do we give them orders? As explained in the introduction, after writing general orders it is usual to write specific orders for each move with simultaneous movement. An alternative system is to have a series of small cards marked, for example, 'form square', 'advance', 'deploy into line'. One of these is placed in front of each

Colonial warfare

unit and when all the cards have been placed, face downwards, they are exposed and movement takes place. If written orders are used and there is an umpire, he should exploit any ambiguities, since mistakes over orders were a not infrequent occurrence in colonial warfare. To cite one actual example, an officer commanding the 3rd Punjab Infantry in the 1868 campaign on the North-West Frontier against the Orakzais was ordered to 'advance to the summit, take the position and hold until further orders'. The officer took this order to mean a large hill beyond the correct one and the Indians were repulsed from a strong position defended by Pathans entrenched in *sangars*. As far as possible, the orders should be similar, in form and substance, to those that would be issued in the field. No verbal communication should be allowed between players on the same side unless the figures which represent them are together on the wargames table.

The extent to which orders will be carried out and how quickly depends on the morale of the units concerned. In view of the different nature of the opposing forces in colonial warfare, the drawing up of morale factors is particularly important in ensuring balanced but accurate games. Account needs to be taken of internal factors, principally the number of casualties suffered by a unit, and of external factors such as the proximity of other units and the nature of the terrain. As in some cases, the Europeans had no option but to fight to the death, the morale effect of losses can only be applied in cases where surrender or retreat would be possible; but it would apply to the tribesmen in almost all cases.

Irregular armies usually contained many waverers and for this reason the critical time in an engagement was often the withdrawal of part of the regular force for tactical reasons, since the tribesmen would gain heart at this apparent sign of weakness and would press home their attack with eagerness and determination. Generally three or four positions in echelon had to be used to cover a retirement and against the Zulus and Afghans a square was usually necessary as these would make short work of any isolated detachments in line. One defeat, even of a small regular force, was sufficient to counteract the morale effect of several previous victories. The loss of artillery, the capture of standards or trophies, or the death of a prominent tribal leader all severely affected the morale of tribesmen. In cases where they were unused to cavalry, even hardened warriors such as the Zulus would be daunted by a determined cavalry charge, particularly by lancers. On the North-West Frontier, the Pathans liked to occupy high ground and to snipe from their positions or charge downhill to mêlée; they thus suffered a deterioration in morale if the regulars 'crowned the heights'. Many irregulars, such as the Boers and the Pathans, disliked having their line of retreat threatened and would retire immediately if this happened. The Pathans' idea of an equal battle was odds of 5:1 or better in their favour so this should be reflected in their morale factors.

Probably the key factor that has to be assessed is the point at which an attacking unit suffers casualties to such an extent that it is unable to sustain the attack. The percentage loss of a defeated enemy in terms of men killed was usually lower than one might expect, for example 20 per cent at Omdurman (Mahdists) and Laing's Nek (British), 17 per cent at Rorke's Drift (Zulus), 9-11 per cent at Abu Klea (Mahdists), Kambula (Zulus) and Gingindhlovu (Zulus) and six per cent at Ulundi (Zulus) and Quintana (Kaffirs). Obviously morale is more important than any other factor in drawing up a set of colonial wargame rules. Probably the best test of whether rules are adequate or not is to play test them using scaled-down representations of one or more of the above battles, using average dice throws, and see if the result you obtain is anything like the actual result.

Skirmish games

A review of colonial wargaming would not be complete without some mention of skirmish games. The essence of this type of game is a time-and-motion study of what a man can do, and how fast, with various weapons. Each figure represents one man and can therefore be given individual characteristics to reproduce his abilities and experience. The ground scale is usually one inch to one yard when using 54 mm or 30 mm figures and 1 cm to 1 yard when using 25 mm figures. One phase represents a short period of time (a second or so). Each man is assessed as a Professional/Veteran, an Average or a Novice and assigned abilities in hand-to-hand fighting and in firing, ranging from 10 for a super-hero to 1 for someone who is virtually useless.

The possible scenarios for skirmish games are as endless as the player's imagination. A few examples* are the death of the Prince Imperial in Zululand, the stand of the last 11 men at Maiwand in the 2nd Afghan War, the charge of the 21st Lancers at Omdurman, the shoot-out at the OK Corral or the legendary fight of the French Foreign Legion against the Mexicans at Camerone in 1863. In my own campaign on the North-West Frontier, which has been in progress for six years now, I use fictitious units such as the Merthyr-Tydfilshire Regiment, the Extra Gurkha Regiment and the 5th Sikh Regiment, all engaged in desperate skirmishes with a villainous Pathan called Bahram Khan and his gang of outlaws. This campaign has provided David Gander, who usually commands the British, and myself, as bandit leader, with a great deal of interest and enjoyment.

To conclude, therefore, colonial wargaming is not the easiest of periods for which to write wargaming rules but it is a glamorous and exciting period which amply repays detailed study. As a start, it is well worth reading further

---

*See also *Skirmish Wargaming* by Donald Featherstone (Patrick Stephens Limited, Cambridge, 1975) and 'Personalised Wargaming: the Seige of Chitral 1895' by the same author in *Battle* for January 1978, page 20.

books about the period, while the occasional exhibition or film may also be of value. Indeed my own interest in colonial wargaming was stimulated by seeing the film *Zulu* some 14 years ago (I actually saw it, for the first time, in Jamaica when the stabbing of each redcoat was greeted with thunderous applause!). Fired with enthusiasm, I began converting Airfix World War 1 figures to British infantry and 'Tarzan' figures to Zulus. Since then, I have graduated to many aspects of colonial wargaming and I can thoroughly recommend the period as one which will repay enormously the time and effort devoted to it.

Rules for Colonial skirmishes

For a variety of reasons, we must make a break here with the 'modular' rules preceding. By and large, wargames fought in this and later periods are predominantly best done—in my own opinion—as skirmishes, in which one model soldier represents one actual man and in which the wargamer is no longer a Brigadier or General but a Private or Sergeant. This is particularly true given the diversity of opposing forces and weapons during the Colonial period and the virtual impossibility of trying to recreate World War 1 battles like the Somme or Ypres with any semblance of realism even if you use a reduced man to figure ratio of 200:1, let alone 20:1 as in the preceding sections. The same applies to World War 2 armoured warfare, wherein one model tank comes to represent one actual tank.

The rules following this and the next chapter are based upon those written by Mike Blake, Ian Colwill and the greatly missed Steve Curtis in Don Featherstone's book *Skirmish Wargaming* (also published by Patrick Stephens Ltd). In these, the following ground rules and sequence of play apply: 1 cm (10 mm) on the table represents one pace; 1 move lasts 5 seconds of real time; and 1 model figure, whether in 25 mm, 30 mm or 54 mm scale, represents one man. The sequence of play is identical to that in the earlier rules except that firing is all done in the 'secondary' stage *after* moving but before mêlée, because the time scale is such that you only get a chance for a single shot anyway.

The model soldiers are each given names and ranks of your choice (eg, Sergeant Sean Courtney and Zulu warrior Mbejane, to cite from one of my favourite authors) and assigned a status which may be either Veteran, Average or Novice; these are pretty self-explanatory and reflect each man's training, experience, morale and leadership ability. Generally, the proportions of each status within opposing groups of figures should be kept similar *or* the numbers on each side varied to give a small group of grizzled veterans a fair fight against a larger number of hairy peasants!

Since the action phase of five seconds is so short, orders for each figure can be reduced to a simple level of 'walk', 'load', 'kneel', 'fire', etc. Only one command may be given per move except that a man with a previously loaded weapon may fire, at reduced accuracy, whilst moving.

There are no distinctions between light, medium and heavy troop types in skirmish games, so the following movement rates apply across the board. Infantry may walk 4 paces per move, run 8 (Zulus 10), charge 10 (Zulus 12) or crawl or limp 2; these factors are halved in difficult terrain, eg, a rocky mountainside, a dense wood or a swamp. Cavalry may walk at 4 paces per move, trot at 8, canter at 12 or gallop at 16. However, they cannot change from a walk to a gallop instantaneously so to accelerate or decelerate take one move to progress up or down one movement band.

All other types of movement, for example, mounting or dismounting, kneeling, diving for cover, drawing a weapon, etc, take one move.

Firing of missile weapons, whether native bows and muskets or European rifles and revolvers, may be either 'aimed' or 'snap'. Aimed fire obviously enhances accuracy, but takes two moves, the first move being to aim and the second to fire, during which time the man may not move or indulge in any other activity. With single-shot weapons the procedure is thus 1) load, 2) aim, 3) fire, 4) reload. Revolvers and repeating rifles do not have to be reloaded until their magazines are empty, thus increasing their rate of fire. 'Snap' shooting with any weapon may only be at revolver ranges.

Calculation of casualties in skirmish games depends on a useful device called percentage dice, as described in the Introduction. To calculate the effect of your firing, establish from the following table the type of target and your percentage chance of a hit at the appropriate range, then add or subtract from the following percentage variations, and finally throw the two percentage dice.

As an example, if you have a 50 per cent chance of a hit, then anything from 0 to 4 on the 'tens' dice and 1 to 9 on the 'units', plus of course a 5 and a 0 respectively, will give you a hit, anything else a miss; ie, a 4 and a 9 (49) is a hit, 5 and a 1 (51) a miss. (It can be most frustrating, too! In one naval encounter I played using percentage dice, my opponent, the Captain of HMS *Hood,* stood a one per cent chance of hitting my own *Bismarck*'s magazine . . . and succeeded! A true reversal of history but I made sure the rules were changed for the next battle so that it could not happen again.)

Chance of hit

		Range	
Type of target	*Short*	*Medium*	*Long*
Standing	80%	70%	50%
Walking or Making an Action	70%	60%	40%
Running or Trotting	60%	50%	30%
Kneeling, Charging or Cantering	50%	40%	20%
Galloping	40%	30%	10%
Lying or In Soft Cover	20%	10%	0%
In Hard Cover	10%	0%	−20%

Percentage variations

Novice shooting	−20%	Running/Cantering or	
Veteran shooting	+20%	Galloping	−20%
Snap firing at Medium range	−10%	Each Light wound	−10%
Snap firing at Long range	−30%	Each Serious wound	−20%
Walk/Trot/Action	−10%		

Once you have scored a hit, refer to the following table whereon the top horizontal line represents the final calculated chance of a hit, the inner vertical columns represent the actual dice score thrown, and the lettered vertical column gives the category of wound.

Casualty table

	1	*10*	*20*	*30*	*40*	*50*	*60*	*70*	*80*	*90*	*99*
A	1	1	1	1	1	2	3	3	4	4	5
B	—	3	7	10	14	17	21	24	28	31	35
C	—	—	8	12	16	20	24	28	32	36	40

Rules for Colonial skirmishes

	1	*10*	*20*	*30*	*40*	*50*	*60*	*70*	*80*	*90*	*99*
D	—	4	9	13	18	22	27	31	36	40	45
E	—	5	11	16	22	27	33	38	44	49	55
F	—	6	12	18	24	30	36	42	48	54	60
G	—	8	16	24	32	40	48	56	64	72	80
H	—	—	17	25	34	42	51	59	68	76	85
I	—	9	18	27	36	45	54	63	72	81	90
J	—	10	20	30	40	50	60	70	80	90	99

To determine the damage you have done, read along the top line to the calculated chance of a hit (say 50 per cent, as in our previous example), then down to the nearest number higher in value than the actual score or, if possible, the actual number (in this case 50 again), then finally across to the vertical lettered column to determine the wound category (in this case J, a light leg wound).

The wound categories are as follows: **A**—dead; **B**—serious body wound—target incapacitated for six moves and can then only crawl; **C**—serious right arm wound—target incapacitated for four moves and may not use that arm again; **D**—as for C but on left arm; **E**—serious right or left leg wound—target incapacitated for four moves, then may only crawl; **F**—light head wound—knocked-out for four moves; **G**—light body wound—target incapacitated for two moves and thereafter all actions at half normal speed; **H**—light right arm wound—target incapacitated for one phase and drops anything in right hand; **I**—as for H but on left arm; **J**—light leg wound—target incapacitated for one move and thereafter cannot run or charge.

Note that all serious wounds knock the target over, as does a light head wound; all subsequent actions take twice as long with any serious wound and the man cannot charge or dive.

When firing at a mounted man, A to F gives a hit on the rider; throw again as if with a 99 per cent chance of a hit to determine wound; G and H kills the horse and throws the rider; I and J wounds the horse, which will automatically slow to a walk for the remainder of the game. If the horse is killed and the rider thrown, there is a 50 per cent chance that he will be injured, so throw the dice again.

Obviously, if a target is protected by hard cover, only wounds on exposed parts of his anatomy are valid; for example, if you can only see the upper part of his body, leg wounds are automatically invalidated.

Mêlées are fought in much the same way as in the 'modular' rules, with factors being assigned to the various weapon types and the chances of inflicting an injury are based on the disparity between them. The weapon factors are as follows: **A**—unarmed; **B**—club or truncheon, blackjack, etc; **C**—knife, dagger or hand-held bayonet; **D**—clubbed rifle, etc; **E**—axe, hatchet or tomahawk; **F**—short sword or sword bayonet; **G**—fixed bayonet, assegai, lance, spear, etc; **H**—long sword; **I**—any cavalry weapon other than those in J and K; **J**—lance or spear; **K**—cavalry sabre or long sword.

Now look at the following table, on which the top horizontal line represents the opponent's weapon, the outer vertical column your own man's weapon, and the inner vertical columns the percentage chances of striking your opponent. Add or subtract from the following percentage variations table then throw a dice and calculate damage using the preceding firing table and wound categories. The only difference is that the mêlée continues if one or both men are only lightly wounded, but they lose a percentage of their effectiveness. If both men score an identical wound, they are assumed to have blocked each other and continue as normal.

		A	B	C	D	E	F	G	H	I	J	K
Own Weapon	A	50	40	40	30	30	20	20	10	20	10	0
	B	60	50	40	40	30	20	20	20	20	10	0
	C	60	60	50	40	30	30	20	20	30	20	10
	D	70	70	60	50	40	40	30	20	30	20	10
	E	70	70	70	60	50	40	30	30	40	30	20
	F	80	80	70	60	60	50	40	30	40	30	20
	G	80	80	80	70	70	60	50	40	40	30	20
	H	90	80	80	80	70	70	60	50	50	40	30
	I	80	80	70	70	70	60	60	50	50	30	20
	J	90	80	80	80	70	70	70	60	70	50	60
	K	100	90	90	80	80	80	80	70	70	40	50

Percentage variations

Type
Veteran v Average	+10%
Veteran v Novice	+20%
Average v Veteran	−10%
Average v Novice	+10%
Novice v Average	−10%
Novice v Veteran	−20%

Movement
Charging	+10%
Opponent Trotting	−10%
Opponent Cantering	−20%
Opponent Galloping	−30%

Miscellaneous
Opponent on left or rear of cavalry	−20%
For each Light wound	−10%
For each Serious wound	−20%
Wounded in previous round	−10%
Opponent wounded in previous round	+20%
Outnumbered, each extra opponent	−10%

Protection
Using Shield	+10%
Opponent has Shield	−10%
Opponent in Cover	−20%

Weapon ranges

Weapon	Range in paces			Time to reload
	Short	Medium	Long	
Bow	50	100	200	1 move
Throwing spear	10	15	25	—
Smoothbore musket	50	100	200	5 moves
Single-shot rifle	50	200	400+	1 move
Magazine rifle	50	200	400+	1 move to insert new loaded magazine; 3 moves to load magazine.
Revolver	10	25	50	3 moves

Chapter 9

World War 1
by Stuart Asquith

World War 1, or the Great War as it was originally known, has for the greater part been bypassed in terms of interest by wargamers. Naval actions, looked at later in this chapter, are perhaps the main exception to this and recently one or two promising sets of rules for air warfare in the period have been available, but not with any marked degree of popularity. The land warfare would seem to be dominated by the fighting on the Western Front and this in turn is influenced by trench warfare—difficult enough to re-enact on the wargame table.

However, there were many other areas of conflict during World War 1. On the Eastern Front, massive Russian and German armies manoeuvred and fought each other in the bleak Polish marshes—here the nature of warfare was more open than on the Western Front and cavalry were used to a much larger extent. The Dardanelles saw British Empire troops trying desperately to establish a bridgehead at Gallipoli under heavy fire from Turkish forces. Palestine, too, was host to some fierce fighting as cavalry led by General Allenby, ably assisted by Arab guerillas under T.E. Lawrence, scored some great successes, culminating with the occupation of Jerusalem in December 1917. Italy also had her share of fighting and many vigorous actions were fought amongst her mountain peaks. To a certain extent Colonial warfare also flourished—German garrisons in East, West and South West Africa had their share of fighting as did similar units in East China, the Caroline and Marshall Islands, Samoa and New Guinea.

Whilst this chapter is not the place to list all actions or even campaigns, the foregoing should serve to underline just how widespread were the areas of fighting.

Weapons

As ever in warfare, weapons developed very quickly during World War 1 and several new ones made their appearance. January 1915 saw poison gas used for the first time in warfare on anything like a large scale. The Russians were the first to suffer from its effects, although in the early stages such gas was really no more than tear gas. As the war dragged on the gases became more and more lethal, attacking not only the respiratory system, but the skin as well. Perhaps fortunately for the Allies on the Western Front, however, the prevailing wind was westerly—a distinct disadvantage to the Germans in their efforts to utilise the new gases.

The allocation of machine-guns to the infantry battalion rose dramatically during the war. At the onset the average German regiment had six machine-guns, at the end it had an average 25. In addition special machine-gun units were also operative to develop the full potential of the weapon.

Hand grenades too were in use. Whilst not a new weapon—they had been employed during the 17th century—their usefulness as a high trajectory infantry weapon was greatly appreciated in the desperate trench warfare of the Western Front. Trench mortars with the same basic capabilities were also speedily developed. Any weapon which could be sited in or near a trench to lob death and destruction into enemy trenches was eagerly snatched up.

The artillery had not in essence changed radically from the Franco-Prussian War period. The French artillery arm was well provided with the famous 'Seventy-Five', a very useful 75 mm field piece—probably superior to the 7.7 cm field gun of the German artillery. The British army had two main field pieces, the horse artillery 13 pdr and the foot artillery 18 pdr, both as good if not better than any other pieces at the start of the war, although outranged by the German gun. Once trench warfare was the order of the day, however, it was found that the then familiar field artillery was of little value. The larger pieces of the foot artillery regiments as a result came into play.

A foretaste of what was to come was seen at Verdun. In February 1914 the German bombardment of the protecting forts was reputably heard 100 miles away, while in a preliminary bombardment before the battle of the Somme in 1915 the Allies fired $1\frac{3}{4}$ million shells at the German lines.

The average range of field artillery pieces had been around the 6,000 yard mark, but those of the Foot artillery were over 10,000 yards. This meant that medium or heavy guns could be sited five miles behind the front line and still cause havoc on enemy positions.

The famous German 'Big Bertha' howitzers had a range of over 15,000 yards (about eight miles). The Austrian-made Skoda 30.5 cm howitzers which played a vital role in crushing Belgian forts also had a range of approximately eight miles—but by 1917, guns with ranges approaching 13 miles were not exceptional. All these figures, however, are dwarfed by that of the 'Long Max' or 'Williams Gun' which shelled Paris from an incredible distance of 67.6 miles.

The last weapon to examine is the one which finally achieved the elusive 'breakthrough' on the Western Front and put an end to trench warfare—the tank. As a result of the Landships Committee set up in February 1915 by Winston Churchill, Britain was able to develop the tank as a weapon.

' Little Willie', the first such vehicle, appeared late in 1915. It weighed 14 tons, had a crew of five (two of them drivers) and a cross-country speed of $1\frac{3}{4}$ mph. From here on things developed apace and by June 1916 the first Mark 1 tanks were delivered for units to begin training with them. Some of the tanks were armed only with machine-guns and were called 'Female' tanks—others had two 6 pdr guns plus three Hotchkiss machine-guns and were termed 'Male'. The Germans were much slower in developing the weapon but after the first appearance of tanks on the Allied side, their progress stepped up.

The very first tank versus tank action happened on April 2 1918 when a Mark IV (Male) and two females met a German A7V tank—regrettably the British tanks came off worst.

The first French tanks appeared in April 1917 a month after the British had used them on the Somme. By the end of the war development had progressed

World War 1

sufficiently for tanks to achieve speeds of up to 20 mph and to take on many of the characteristics seen in their modern counterparts.

The men

Lastly, what of the men who took part in the battles—initially in the open and subsequently in the trenches?

The French entered the war confident of an early victory. Despite his predecessor's defeat in the Franco-Prussian War of 1870, the average French soldier felt he was more than a match for the hated 'Hun'. The French uniform too was reminiscent of an earlier war. The long blue coat and baggy red trousers were not the best clothing for the war that was to be. The authorities realised this soon enough and in the winter of 1914 horizon blue uniforms were issued and a similarly coloured helmet replaced the blue and red képi.

The British Army arrived at a strength of two Corps in 1914 under the name of the British Expeditionary Force (BEF), a name it retained throughout the war, in spite of the fact it grew in size and many of its original men did not survive the awful battle of the Somme.

Probably the best-equipped and best-trained army Britain has ever put into the field, the BEF stunned the Germans with its rapid rifle fire. Contemporary German sources reveal they were convinced that all the BEF were equipped with light machine-guns after the initial clashes, so fast and accurate was the British firepower.

The German soldier moved into the war as a tool of the much doctored Schlieffen Plan, which called for a cautious central advance—reversals were anticipated—and a devastating 'right hook' sweeping around Paris. The BEF supported, intermittently, by the French armies, interfered with this plan at Mons and stopped it dead at the River Marne—so trench warfare was born.

The German infantryman was well equiped, well catered for and well led—the main weakness if any seems to have been much higher up the pecking order.

Photograph taken by a reader of Airfix Magazine *which represents a superb late war encounter between British 'Male' and 'Female' tanks accompanied by infantry, and German troops supported by a solitary A7V.*

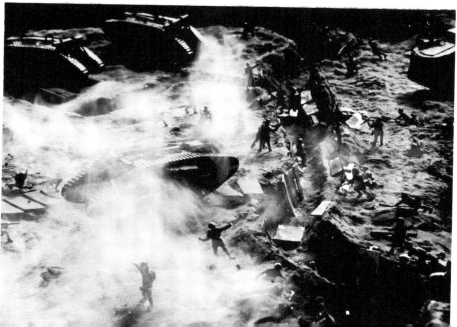

Above *A view of an early First World War encounter with the troops still drawn up in essentially 19th century formation (*Michael Turner*).*

Left *Close-up of a Peter Laing 15 mm scale German machine-gun and crew (*Alan Wright*).*

A brief mention only needs be made of the cavalry of all nations on the Western Front. Whilst warfare was still open in the early months of the war, cavalry had a role to play. Gradually, as the trench lines extended in length and depth, the cavalry were kept in readiness to exploit the breakthrough that never came. Before long they took their places in the line and fought alongside their infantry colleagues in the trenches.

Naval warfare

At the commencement of hostilities in 1914, Great Britain had the strongest Navy, both in terms of power and numbers, in Europe. Due to some commandeering of vessels being built in England for foreign Navies, Britain had 24 'Dreadnoughts' available for service compared to the 13 of Germany. This was also true of the battlecruiser, the Royal Navy having nine such vessels available, the Germans six, some of which were inferior in gunnery. 'Dreadnought' is a word used to describe battleships built after the launching of HMS *Dreadnought* in 1906.

The vessel's completion with ten 12-in guns rendered all previous battleships ineffective. All major powers had to scrap their building programmes and start afresh. From this time on, naval power was reckoned in units of 'Dreadnought' battleships and battlecruisers.

As regards the other nations involved, France led the field with capital ships followed by Russia, Austria-Hungary and Italy. One more nation has to be considered, however, when one reviews naval potential at the onset of World War 1. Japan had been an ally of Great Britain since a treaty in 1902. This did not obligate her to join the struggle on the side of the Royal Navy—but seeing the opportunity the war offered for expansion, Japan did just that.

When war came then, Britain and her allies were seemingly numerically and technically superior to Germany, Austria-Hungary and Turkey, and geographically well-placed to interfere with any of their strategic plans. However, things were not always what they seem. In summary the British ships lacked the wide beam of their German counterparts due mainly to the limitations imposed by available dock facilities. Due to cut-backs in spending, Royal Navy ships had had to be constructed to dimensions that would fit into existing dockyards rather than the reverse. These strictures did not apply to the Germans and, as a result, their wide beams made them less susceptible to mines and torpedoes and meant they could carry wider armour for protection.

Secondly, German capital ships had been built to different criteria to those of the Royal Navy. Great Britain was a global power—her capital ships carried enough coal for an average operational radius of 4–5,000 miles, the crews living permanently on board. This was not the case with the German ships—their crews lived ashore in barracks when the vessels were in harbour and only manned the ships for the duration of a specific and usually localised operation.

The end product, brought about by the saving in fuel weight and crew quarters, meant the German ships could have many more watertight compartments below decks and were virtually unsinkable. Gunnery ability in terms of both performance and weight were about equal between the two Navies but the German gunnery control system was marginally better. In the areas of shells, mines and torpedoes, too, the Germans were several steps ahead of their British counterparts. So it can be seen that, overall, there was little to choose between the two Navies at the outbreak of war in 1914.

Without going into too much technical minutae, one or two comparative remarks on the weaponry and defences of 'Dreadnoughts' may be of interest to the potential wargamer.

The British 'Super-Dreadnought' HMS *Queen Elizabeth* displaced 33,000 tons, was 640 ft long and developed a speed of 25 knots. Her main guns were eight 15-in supported by 12 6-in and 12 12 pdrs. Her 1,016 crewmen were protected by a 13-in armour belt and 11-in of armour on the gun turrets. As a direct comparison, SMS *Kaiser,* a German 'Dreadnought', displaced 24,380 tons, had a crew of 1,178 men and a $13\frac{3}{4}$-in belt with $11\frac{3}{4}$-in turret armour. Developing a speed of 23 knots, the *Kaiser* had ten 12-in, 14 5.9-in and eight 3.4-in guns.

Naval engagements were in reality few and far between during World War 1 and there follows a few notes on the better known actions.

The Cruise of the *Emden*—In August 1914 the German light cruiser *Emden* (ten 4.1-in guns, 3,600 tons) sailed into the Indian Ocean—until then a veritable British 'lake'—and for three months sank shipping and inflicted some £15 million worth of havoc. HMAS *Sydney* (5,600 tons, eight 6-in guns), a protected cruiser, finally caught up with the *Emden* and in a savage gunnery engagement forced the German Captain to run aground and surrender.

The Falkland Islands (December 1914)—A British cruiser force comprising *Inflexible* (battlecruiser), *Invincible* (BC), *Carnarvon* (armoured cruiser), *Kent* (AC), *Cornwall* (AC), and *Glasgow* (Light cruiser) clashed with a similar force of German ships, *Gneisenau* (AC), *Scharnhorst* (AC), *Dresden, Leipzig* and *Nurnberg* (LC), under the command of Admiral Von Spee. It was an overwhelming victory for the Royal Navy, who sank four out of the five German vessels. More importantly, perhaps, it made the German High Command decide not to risk its High Seas Fleet against the British Navy but to return it to the area of the Baltic.

Dogger Bank (January 1915)—Another clash between opposing battlecruiser squadrons, the action at Dogger Bank was a running fight with the Royal Navy chasing the Germans. The latter lost one vessel sunk *(Blucher)*, one damaged *(Seydlitz)* and the British flagship *(Lion)* was badly damaged.

Jutland (May 31–June 1 1916)—The major naval engagement of the war and in the final analysis a draw. A German battlecruiser force was sighted and engaged. The action escalated from then on and brought in the German High Seas Fleet and the British Grand Fleet. Small actions within the main battle were the order of the day and in the early hours of June 1, the British forces disengaged to avoid the risk of running into enemy minefields. Thus ended the first and only great naval gun battle between ships of the Dreadnought era.

The British Grand Fleet had greater losses than the German High Seas Fleet but victory, if any, went to the Royal Navy since the Germans did not again venture out seeking conclusions.

Naval wargaming is one aspect of World War 1 which is popular and rules* and ships in various scales are plentiful. Three main types of wargames present themselves as choices.

Firstly, hunt the raider—a single ship (eg, the *Emden*) causes havoc against merchantmen and has to be hunted and killed by a pursuit force. Secondly, a battlecruiser action such as the Falkland Islands or Dogger Bank means the wargamer can have some heavy guns in play, but need not spend out on massive fleets. Finally, the grand slam—the 'Jutland' type action. Whilst the opposing fleets can be recreated ship for ship (details of all involved are easy enough to come by) the concept is not as satisfying as it may initially appear. The average playing area for naval wargames, be it tabletop or floor, is simply not big enough for accurate scale distances to be achieved. Squadrons finish up cramped in corners and the whole set-up generally declines to a frustrating, premature end.

As stated, World War 1 naval wargaming is very popular and the wargamer wishing to explore this aspect of the period will find information readily available along with all the paraphenalia necessary to recreate the engagements.

Air power

Air power was virtually non-existent at the outbreak of World War 1. None of the major nations had shown much interest in the aeroplane as a weapon and little had been done—in the military sense—since Blériot flew the English Channel in 1909. The British General Staff did not even consider the aeroplane for reconnaissance purposes, deciding that cavalry were far more suitable for the job.

In France the lower echelon staff were keen enough on the idea but the Army and Navy brasshats were not impressed. Even though the French Army did indeed have an air arm by 1914, the whole affair was somewhat nonchalant.

The Germans had been more far-sighted and had organised the infantry to feed off reconnaissance reports from aeroplanes. The main German interest at

---

*See in particular Paul Hague's *Sea Battles in Miniature*, published at the same time as this volume by Patrick Stephens Ltd.

this time was in airships, however—bolstered mainly by support from the German Naval Air Service. Large rigid airships had a tremendous range and cargo capacity and it was mainly for these reasons that the development programme continued. By their very nature, of course, both sheer size and packed full of hydrogen, these airships were somewhat vulnerable when utilised as a military weapon.

All the foregoing, however, is not to say the major powers had no aircraft—they did and in reasonable quantity—but it seems they were unsure exactly how to use this new arm. Flying schools were established and as war loomed it seemed the best use of aircraft was reconnaissance, pure and simple. Naturally enough the next logical step was not only to have a look at the other fellow's deployments but to prevent him seeing your own. At about the same time, the thought came to light, that whilst cruising over enemy installations, ie, targets, it might not be a bad idea to drop something nasty over the side. So it was that war gained a third dimension and shifted for the first time into the air (although the Italians had earlier used a Blériot to reconnoitre Turkish dispositions in 1911).

Belgium, Austria and Turkey had little to offer in either numbers or ideas in the way of military aviation. Russia was at about the same level but details of the Russian early days of aircraft development are hazy.

Gradually the aeroplane began to gain some credibility in the minds of the various General Staff as a useful reconnaissance tool but still no one seemed to realise its offensive potential. Tactics of air warfare had not even been considered and the question of arming aircraft had not been seriously explored.

Whilst this is not the place to enter into reams of technical data, a few remarks on the performance of the aeroplanes of 1914 may assist the reader in putting events into perspective. The Bristol Box Kite was the first British mass-produced aircraft and was used largely for training purposes. It was a heavily braced pusher (ie, engine pointing to rear) biplane, having a 50 hp engine and a maximum speed of 40 mph.

The French Blériot XI—a militarised form of the Channel flight aeroplane—had a 70 hp engine and could clock up a maximum speed of 66 mph at sea level.

The Etrich Taube (Dove) built by an Austrian engineer had a larger 100 hp engine and could fly at over 71 mph, again at sea level. The German Rumpler could reach speeds in excess of 90 mph but it was powered by a larger 160 hp engine.

These very few examples selected virtually at random from a list of aircraft flying in the first few weeks of the war serve to underline the low speeds which limited their performance.

Gradually as the war progressed, pilots and observers, if carried, took to shooting at targets and available enemy aircraft. This led inevitably to the next step—the fixing of machine-guns to the planes in order to mount 'offensive patrols'. This was taken a stage further by the development of what is usually termed the world's first fighter, the French Morane-Saulnier N. Fitted with deflector plates on the propeller, the system allowed the pilot to fire his guns whilst seated—previously he had had to stand up to do this—and deflected any bullets away from the spinning propeller. The Germans improved on this idea and before long all new fighters were being fitted with synchronising equipment to fire the guns so as not to hit the propeller at all.

War has ever stepped up technological advances and in keeping with this new and better machine-guns become available for plane designers to use. Tactics emerged and dog-fights dotted the sky over the trenches of the Western front. Artillery spotting was developed into a fine art, but the lumbering two-seaters were easy prey for marauding fighters and so had to be defended by friendly counterparts.

Squadron sizes and composition altered and changed from nation to nation and it would be tedious to list all the alterations here. One point the wargamer should bear in mind, however, is that certainly in the early part of the war, squadrons would have several types of aircraft in their ranks. Gradually a consistency of sorts came in but it was not until after hostilities ceased that aircraft types were sorted within the same unit.

By the end of the war aircraft production had increased in leaps and bounds. In July 1918 the Germans produced nearly 1,500 new aircraft—but the problem by this time was the dwindling numbers of pilots to fly the new machines. Air superiority was the province of the Allies and in the last few months of the conflict troops, trains, transport and supply dumps all came under attack.

A passing mention should perhaps be made regarding one or two other aspects of aircraft development during World War 1.

The navies of the various countries had seen the potential in the aeroplane as a 'seeking' weapon to be the eyes of the Fleet. Gradually planes came to be launched from the top of gun turrets of battleships, only to have to ditch in the sea after their mission. The natural progression to the aircraft carrier comes too late, however, for its inclusion in any study of World War 1.

Bombers had also been developed, usually a few steps behind fighter progress. A natural advance from the slow-moving artillery observation plane, the bomber by 1918 had developed into an important weapon—further enabling the war to be taken behind the lines and into the interior of an enemy country.

So the sky too, now had its share of war. From simple beginnings, the major nations built up huge air forces and the struggle aloft was every bit as intense as the battle on land. The war in the air, however still retained a certain élan and captured the imagination. Oswald Boelcke, Albert Ball, Micky Mannock, James McCudden, Captain Baron Manfred von Richtofen, Max Immelman—the list is long. The names of the 'aces' are still largely remembered today for their part in furthering war in its third and most romantic dimension, the air.

Rules for 'No man's land' skirmishes

Essentially these are the same as for the Colonial period, with a couple of additions. All troops are assumed to be equipped with magazine rifles containing five shots, or six-shot revolvers. Reloading times as before. Light head wounds are invalid on anyone wearing a steel helmet. Entrenching tools with sharpened edges (a favourite weapon of the trench raiders) count as category F in the mêlée table. Climbing into or out of a trench takes one move. Crawling through barbed wire with cutters is at the rate of half a pace per move.

An additional weapon to be considered is the grenade. Throwing range as for throwing spear. Veterans will automatically hit the desired target, Averages have an 80 per cent chance, Novices 60 per cent. A grenade explodes on the phase thrown, and has a blast diameter of 8 paces. Throw for effect on any figures in blast circle as for a 99 per cent chance of a hit.

Chapter 10

Armoured warfare 1939-45
by Bruce Quarrie

Military developments between the Armistice in 1918 and Germany's invasion of Poland which sparked off the Second World War on September 3 1939 were greater than over any previous historical period. A Greek hoplite could have understood a 17th century pikeman; a mediaeval longbowman would have been at home with a Napoleonic rifleman; an English Civil War dragoon would have experienced little difficulty in the American Civil War two hundred years later. But neither they, nor the veterans of the Somme, could have visualised the fast-moving, heavily armed and heavily armoured tanks; the 400 mph cannon-armed fighters; the awesome aircraft carriers; or the atom bombs of the Second World War.

In the context of this book, the most significant development was the tank. Aerial and naval wargaming are beyond our space limitations, requiring entire books to themselves*.

During the period 1918-1939, the tank became an increasingly sophisticated machine when compared with the lumbering monsters which helped to break the stalemate of trench warfare after 1916.

In the Foreword to the excellent booklet *Fire and Movement* published by the Royal Armoured Corps Museum at Bovington in Dorset (which itself is well worth a visit), the problem of tank design is succinctly put as follows:

'A tank is an armour-protected vehicle which carries direct-firing offensive armament to selected positions on the battlefield from which it will have the best effect. The three main characteristics of a tank are therefore firepower, mobility and protection. Their order of priority is a matter for the General Staff to decide.

'An ideal tank would consequently be suitable for any tactical role if it had armament capable of dealing with all varieties of enemy targets, protection to render itself immune from all forms of attack, and the mobility to permit travel at speed for long distances regardless of terrain. Two other criteria are inferred, that the tank must be reliable, and that it must afford a maximum lack of crew fatigue.'

The tanks which saw action at Cambrai in 1917 lacked all of these features: they were slow and mechanically unreliable; they afforded maximum crew fatigue; and they were inadequately armed and armoured.

---

*Air Battles in Miniature: A wargamers' guide to aerial combat 1939-1945, by Mike Spick; and *Sea Battles in Miniature: A guide to naval wargaming*, by Paul Hague, are both available from the publishers of this book, Patrick Stephens Limited.

The inter-war period

During the inter-war years these problems were discussed in America, Britain, France and Russia. They were discussed in Germany too, but Germany was forbidden to build tanks. The result was a wide variety of unsatisfactory vehicles. The British evolved three basic types: the Light, Cruiser and Infantry tanks. Light tanks were small, lightly armoured, equipped with machine-guns, and fast. They were totally unsuited to fight against anything else than other light tanks. Their primary role, however, was seen as the old horsed cavalry one of reconnaissance.

Cruiser tanks were larger and also fast, but they were still lightly armoured and, until later in the war, were only armed with 2 pdr guns. These guns could only fire solid armour-piercing shot; they were incapable of firing high explosive shells and thus, to revert to the above quote, were incapable 'of dealing with all varieties of enemy targets'.

Infantry tanks, as the name implies, were designed specifically as infantry support weapons. They were heavily armoured but extremely slow-moving. As a result of the weight penalty they suffered, initially they too only mounted machine-guns. So, you had one type of tank which was capable of receiving punishment but not dishing it out; one which was capable in inflicting a limited amount and sustaining an equally limited amount; and one which was capable of neither. This state of affairs remained substantially unchanged right through to Dunkirk.

Other countries adopted different approaches, but the concept of three different vehicles for three different tasks rather than a single versatile main battle tank predominated. Developments were most slow in America, which had reverted to isolationism, despite the brilliance of their foremost designer, Walter Christie, whose ideas were adopted with great success by the Soviet Union. In 1940 America literally did not possess a single effective tank.

France wasted much time in the tank 'race' by trying to modernise old First World War designs rather than developing new vehicles. By 1940, however, they had surprisingly developed some of the best tanks in the world, particularly the heavily armed *and* armoured Char B infantry tank and the lighter Renault and Hotchkiss cavalry tanks. Why these vehicles failed against the German Panzer divisions in 1940 was more due to faulty deployment than any design failure.

Some of the most interesting developments took place in Soviet Russia, where the first really effective tanks in the modern sense were created. They suffered their share of mistakes but with one radical difference compared to France or Britain: instead of concentrating on protection or speed, they concentrated on firepower under the correct assumption that if you could eliminate your opponent's tank before he could get within effective range of his own armament, then protection was less vital. However, the Russians also appreciated the virtues of Christie's ideas on suspension, and managed to combine firepower with good mobility and a reasonable degree of protection in their BT-5 and -7 designs and, of course, the later incomparable T-34. They had their throwbacks—monstrous multi-turreted heavy tanks like land battleships and ineffective light tanks too—but they also developed the world's most effective heavy tank in the shape of the KV-1.

Germany, as stated, was forbidden under the terms of the Treaty of Versailles to build any tanks at all. With Hitler's rise to power all that changed, particularly under the inspiration and determination of Heinz Guderian.

Armoured warfare 1939-45

Guderian had made a detailed study of tank developments in other countries during the 1920s and early 1930s and had come to the conclusion that the whole concept of armoured warfare currently prevailing was radically wrong. Inspired by the writings of the British theoreticians Liddel-Hart and Fuller (who were, inevitably, ignored in their land of birth), he and others of similar mind evolved a whole new system of warfare. And in 1940 the Allies were to shudder at the horrific efficiency of *Blitzkrieg*.

1939-40

Blitzkrieg (literally, 'lightning war') was unique in that, for the first time, it ignored flanks. Flanks, thought Guderian and others like him—such as Erwin Rommel—were redundant. Forget rolling up your opponent's field army: strike for his headquarters and lines of communication; decapitate him rather than amputate limbs.

It made a great deal of sense. Poland was one country to overrun, regardless of strategy; France, the Low Countries, and the British Expeditionary Force were something else entirely. They had numerical superiority. They had more tanks, more infantry, and more anti-tank weapons. But their governing strategies were divisive, and they clung to the concept of the tank as an adjunct to, rather than a decisive weapon for, victory.

There is a tendency amongst military thinkers to prepare for the last war rather than the next; this has amply been demonstrated earlier in this book: for example, Napoleon's enemies in 1805-6 were preparing for an action replay of the Seven Years' War. Similarly, in 1939-40 the Allies were all set for a re-run of 1915. Technological developments meant little to the Colonels of old and hitherto victorious regiments.

In 1939-40, Germany had quite an arsenal of different tank designs. The smallest was the PzKpfw I (PzKpfw = *PanzerKampfwagen* = armoured fighting vehicle, the standard German abbreviation for 'tank'). Originally designed as a training vehicle, it was lightly armoured and carried two machine-guns. The slightly larger PzKpfw II which was designed to supplement it was equipped with a quick-firing 20 mm cannon and a machine-gun.

Next up the scale came the PzKpfw III, a 15-ton tank mounting a version of the Army's standard 3.7 cm anti-tank gun in addition to two machine-guns. Finally there was the PzKpfw IV, an 18-20-ton vehicle equipped with a low-velocity 7.5 cm gun. In addition, the Germans utilised large numbers of the Czech 35(t) and 38(t) designs which, while as light as the PzKpfw II (circa 10 tons), had better armour protection and mounted 3.7 cm cannon.

However, the Panzer Divisions which were deployed with such staggering success during the first three years of the war did not consist solely of tanks, and herein lay the German Army's great advantage. Instead, the Panzer Divisions resembled the old Napoleonic Corps, being formations of all arms acting interdependently—tanks, motorised infantry, self-propelled or towed artillery, motor cycle and armoured car reconnaissance formations and integral signals, anti-tank, medical, supply and workshop units.

In other countries the tendency was to keep the tanks separate from the infantry except when required for specific support tasks. The result was that, in 1939-40, although possessing overall superiority in numbers, the Allies were unable to muster their tanks in sufficient concentration to beat back the fast-moving Panzers. Even so, British Matilda Infantry tanks with their very thick

The photographs in this chapter all illustrate 1:76 or 1:72 scale tanks and figures although, as explained, these are not nowadays often used in World War 2 wargames as the 1:285 or 1:300 scale models are so much more suitable. However, the latter are almost impossible to photograph! This picture shows my ruined farmhouse again with a 1940-period French Char B (scratch-built) with an Airfix Universal Carrier and a Light Tank Mark IV converted from the latter.

armour administered a check to Rommel's Panzers at Arras and were only defeated when he brought forward several dual-purpose 8.8 cm anti-aircraft/anti-tank guns.

After the debacle or miracle of Dunkirk, depending on which way you look at it, the German Army had control of virtually all Europe. Enviously, the Italian dictator Mussolini now decided that he wanted to carve out his own empire in the Mediterranean. In June 1940, therefore, he declared war on Great Britain and Italian troops invaded Egypt. To begin with, the British troops fell back before them, but in December General Wavell launched a counter-offensive which rapidly broke up the Italian forces and sent them into headlong retreat. Within ten weeks the Italian army had virtually ceased to exist and the Commonwealth forces were poised to seize Tripoli itself.

However, in October 1940 Italy had also invaded Greece, and British support was urgently required there. As a result, the desert force was denuded of many of its best formations, as the British Isles were still preparing for a German cross-Channel invasion and could spare little help for the Mediterranean theatre. But the Italians fared as badly in Greece as they had in North Africa, and if Hitler did not want to see his dubious ally go under, he had to send assistance.

1941

1941 proved a year of disaster for the Commonwealth troops in the Mediterranean. Greece fell to the Germans, then the strategically important island of Crete to a large-scale airborne assault. In North Africa, Rommel arrived with the first elements of what was later to become the famous Deutsches Afrika Korps. He brought with him substantial numbers of the PzKpfw II, III and IV tanks, and immediately began planning a counter-offensive. The attack opened at the end of March and by the middle of April the British had been pushed back into Egypt, with the exception of a stubborn garrison in Tobruk.

A scene from a Western Desert wargame. British Matilda 2s supported by a 25 pdr gun are threatened by three PzKpfw IVF1s.

At this stage the British had few tanks in the desert capable of standing up to their German opponents, despite their succcess over the poorly equipped Italians. The best tank was the Matilda II with 2 pdr gun and very thick armour, but it only had a cross-country speed of 8 mph. In addition, they had some Valentine heavy tanks which at this time were comparable to the Matilda, although later versions were up-gunned with 6 pdrs and 75 mm anti-tank guns, making them more of a match for the German Panzers.

There were a number of machine-gun armed Mk VI Light Tanks, about the equivalent of the German PzKpfw I; obsolete A9 and A10 Cruiser Tanks and the improved A13 ; and the famous Crusader (A15), a very fast Cruiser Tank although lightly armoured and, again, only mounting a 2 pdr gun.

However, both the Crusader and the Valentine possessed turret rings of sufficient diameter that their turrets could be fitted with larger weapons, and both went through the same progression from 2 pdr to 6 pdr to 75 mm.

Turning now to the east, in June 1941 Hitler unleashed Operation *Barbarossa*— the ill-fated invasion of Russia which was to lose him the war. Like Napoleon, he had assembled the mightiest force he could muster, and to begin with the invasion looked like a repeat of the sweeping Blitzkrieg successes of the previous two years. The Russian army was disorganised and suffered an acute lack of competent generals due to Stalin's earlier purges. To begin with, they too fell back reeling in disorder, and the three German Army Groups swept inexorably closer towards their initial objectives of Leningrad, Smolensk and Kiev. Russian troops, tanks and guns were captured in their hundreds of thousands in massive encirclements.

Now, however, Hitler made a fatal mistake. Instead of concentrating on the prompt capture of Moscow—which, far more so than in 1812, was a vital communications centre—he split his strong central force to give assistance to his northern and southern flanks. Although these operations were successful, and resulted in the capture of Kiev and containment of Leningrad, vital weeks had been lost. By the time the order was given to resume the advance on Moscow, autumn lay upon the country and the Russians had been given the opportunity to prepare elaborate defences around the capital.

The autumn rains produced thick mud in which fully tracked vehicles floundered, while all other forms of transport were virtually immobilised. The

A Russian scene from later in the war; visible are two Tiger Is, two PzKpfw IVs, SdKfz 7 and 251 half-tracks, a Nashorn self-propelled gun and an SdKfz 232 armoured car

Russians were in better shape as most of their tanks were fitted with comparatively wide tracks which spread their weight better, enabling them to keep up a semblance of normal movement.

The most important Russian tank at this time and throughout the war was the T-34, a vehicle which has often, and justifiably, been acclaimed as the best tank of all time. Certainly its continued front-line use in many countries throughout the world nearly 40 years later is vindication of this. The early T-34/76A was classed as a medium tank although in 1941 it weighed more than the PzKpfw IV variants in service, at 26 tons. It had good armour protection—up to 45 mm in thickness compared with the PzKpfw IV's 30 mm—and mobility (up to 31 mph), and was armed with a 76.2 mm anti-tank gun capable of firing armour-piercing shot or high explosive shells. This combination was superior to anything the Germans possessed until the introduction of the PzKpfw IVF2 in 1942, up-armoured to 50 mm and with a long-barrelled 7.5 cm gun.

In addition to the T-34, the Russian armoured forces deployed a variety of light tanks, some being amphibious and armed only with machine-guns, others being equipped with 45 mm guns which were powerful for tanks of their size. They also had the BT-5 and -7 medium tanks which were comparable to the PzKpfw III, although the BT-7 was better armed with a 76.2 mm gun; the multi-turreted T-28 and T-35 vehicles which proved a tactical disaster; and the KV-I heavy tank which carried the same gun as in the T-34 but which had armour plate up to 110 mm in thickness.

Thus, faced with a combination of stiffening Russian resistance and the beginning of one of the coldest winters on record, the German advance on Moscow slowed and eventually stopped, even though leading elements of the Panzer Divisions had penetrated into the city's suburbs.

Events on the far side of the world now came into play; the Japanese attacked the American naval base of Pearl Harbor and the United States was dragged unwillingly into the war. Meanwhile, in North Africa, a British offensive designed to relieve Tobruk had failed in the summer of 1941, but a renewed attempt in

November came as a surprise to Rommel and was successful despite a bitter struggle, and the Afrika Korps was forced to retreat. Some of the fiercest tank battles of the campaign took place during this period, and the balance of power swung backwards and forwards like a pendulum. However, for the first time Rommel found himself seriously outnumbered as the Commonwealth forces had been substantially reinforced, not least with the new American M3 Stuart light tank, nicknamed 'Honey' by its crews.

1942

As the war entered 1942 the situation looked grim for the Allies despite the local success in North Africa. In Russia, the Germans contained a Soviet winter offensive although they were forced to give some ground, then began preparing for their next summer campaign. Moscow had been abandoned as an immediate objective. Hitler now had his sights set on the rich oilfields of the far Caucasus, and attention switched to the south.

As in the opening days of the campaign, to begin with the Panzer divisions were everywhere victorious; but the Russians were trading ground—of which they had plenty—for time, time in which their munitions plants could get into full production. German forces swept into the Crimea and invested the important naval base of Sevastopol; they pushed vigorously across the seemingly endless steppes, where heat and dust now became their chief enemies. They reached and crossed the River Don and entered Stalingrad; and they reached the Caucasus oilfields, only to find that they had been rendered useless by Soviet demolition teams. Then they were checked. The Russians launched a winter counter-offensive. The bitter fighting for control of Stalingrad continued, even when von Paulus' VI Army was cut off, until the frostbitten, weary and half-starved German troops were finally forced to capitulate in January 1943.

In North Africa Rommel again seized the initiative in January 1942 and pushed eastward once more. Tobruk was invested for the second time and finally fell to the Germans in June. Despite courageous fighting, the Allied forces were pushed back and back, into Egypt, and finally to the defensive line prepared around the otherwise insignificant desert railway station of El Alamein. If they could not hold here, Cairo and Alexandria were doomed. But hold they did. Rommel tried to break the position during the first battle of Alamein in July–August, but was checked and then forced on the defensive by strong counter-attacks.

At this point Churchill appointed a new commander to 8th Army—Lieutenant-General Bernard Montgomery. 'Monty' was a perfectionist although not as dashing or innovative in his outlook as Rommel, and he clearly saw that a decisive victory in North Africa could only come about after sufficient superiority in men, tanks and guns had been achieved that even superior tactics would be of no avail against them. He also knew, of course, about the impending amphibious assault in French North Africa, behind the Afrika Korps.

Allied forces in North Africa had been receiving quantities of the American M3 Lee and Grant medium tanks during 1942. Although not a brilliant design, with the main 75 mm gun in a side sponson rather than a fully traversing turret, these vehicles at least gave British tanks the ability to fire high explosive shell and fight at greater parity with the latest German PzKpfw IIIs, which were armed with long-barrelled 5 cm guns, and PzKpfw IVs with long-barrelled 7.5 cm weapons. By the time the second battle of Alamein opened in October 1942, the Commonwealth forces had also begun to receive supplies of the new M4 Sherman medium tank, which had a suspension and chassis virtually identical to the M3 but with a revised hull and a revolving turret carrying a 75 mm gun.

The second battle of Alamein opened on the night of October 23 1942 with an artillery barrage which could be heard in Cairo. Then the tanks rolled forward. Minefields proved a serious obstacle and the Germans fought grimly, but they were exhausted, their tanks severely weakened in numbers, while their Italian allies were becoming increasingly reluctant about Mussolini's whole empire-building venture. The result was an inevitable victory for Montgomery, and the Afrika Korps began its long retreat. Then, in November, Anglo-American forces landed in Morocco and Algeria in Operation *Torch*. Squeezed between two foes, the Afrika Korps fought back doggedly and were reinforced by a fresh Panzer division in addition to several of the PzKpfw VI Tiger I heavy tanks which had first seen action earlier in Russia. These very heavily armoured tanks, equipped with a modified version of the infamous '88' 8.8 cm dual-purpose gun, had been produced as the answer to Soviet heavy tanks which, by 1942, were being fitted with 85 mm weapons. Although lumbering and rather unmanoeuvrable, these vehicles were far superior in firepower and protection to any Anglo-American designs, and created havoc wherever they appeared.

Although the untried American troops suffered a severe setback at the battle of Kasserine Pass, the end was in sight for the valiant Afrika Korps. Caught between two pincers, they were forced into an every-decreasing perimeter in Tunisia, and finally capitulated in May 1943.

Tanks were little used in the Far East because the nature of the jungle terrain was inhibiting. This was principally an air-sea campaign. To begin with, and principally due to the element of surprise and Allied unpreparedness, the Japanese encountered little significant resistance. The Phillippines and Dutch East Indies fell, as did Malaya and the British base at Singapore. However, after the crucial aircraft carrier battle of Midway in June 1942, the Japanese were forced increasingly on the defensive. The Americans gradually resumed the initiative in the Pacific, retaking Guadalcanal after a bitter struggle during August 1942 to February 1943, then Papua and New Guinea, through the Gilbert, Marshall and Mariana islands during 1943-44 to Iwo Jima and Okinawa in 1945. Meanwhile, Commonwealth forces on the mainland struggled first to hold and then to resume the offensive in Burma and China during 1943-44. A planned campaign to liberate Malaya in 1945 never materialised because the dropping of atomic bombs on Hiroshima and Nagasaki brought the war in the Far East to a sudden close.

1943

Returning to the Russian front, the Soviets pursued their advantage during the winter of 1943, pushing the Germans back along the whole of the southern front, and recapturing Kharkov. This important city was then retaken by the SS Panzer Korps and the line stabilised while both sides paused for breath.

As spring 1943 turned into summer, the German forces in Russia were in surprisingly good shape, despite the losses of the winter. Planning immediately began for yet another summer offensive, and the little town of Kursk soon found itself at the centre of the biggest tank battle of the entire war. The Soviets had established a large salient around Kursk, and the Germans planned a two-pronged assault from north and south to cut off the 'neck' and trap the Russian divisions in another encircling exercise. However, the Russians also appreciated the significance of the salient, and while the Germans were assembling their forces for the attack, they were also reinforcing and establishing strong defensive positions and tank 'killing grounds'.

The battle of Kursk in July saw the debut of the German PzKpfw V Panther tank, an advanced medium tank which incorporated many lessons learned from study of the Russian T-34. It featured thick and well-sloped armour plate which was capable of deflecting most anti-tank projectiles, and carried a superb long-barrelled 7.5 cm gun. However, in its earliest version it suffered from several defects and proved something of a disappointment.

The battle of Kursk was really the beginning of the end for Nazi Germany. The Russian defences proved stronger than anticipated and, although some headway was made, particularly in the south where a massive battle involving over a thousand tanks on each side developed, the attack was finally called off by Hitler. The reason was the Allied landings on Sicily, threatening the 'soft underbelly' of Axis Europe.

From July 1943 onwards it was apparent to all but Hitler and his sycophants that the war was lost. In Russia the communist forces followed up the victory at Kursk by a determined offensive which, although halted locally on several occasions, followed through, driving the Axis troops ever westward, out of Russia, into Poland and Hungary, and ultimately into Berlin itself. The Allied landings in Sicily proved a mere prelude to amphibious assaults on mainland Italy, where beacheads at Salerno and Anzio were carved in blood. Monte Cassino was finally taken and the Italians capitulated, although fascist elements continued to fight on. Rome fell and the Allies advanced upon the mountainous defences of the Gothic Line.

1944-45

Then, on the glorious sixth of June 1944, the long-awaited 'second front' opened. American, British, Canadian, Free French and Polish troops and other Allied contingents swarmed ashore on the Normandy D-Day beaches. The disbelieving German high command tried to rally and respond, but to little avail. The Allies had their setbacks but, once ashore, they refused to give up. Aided by an aerial supremacy whose effect has often been overrated, they finally broke out of the stumbling blocks of Caen and St Lô, and trapped the pride of the German Panzer divisions in the Falaise pocket. Paris was liberated and the Germans were forced back wherever they were encountered.

By this time several new tanks had made their debut in both the German and the Allied armoured divisions. Britain's earlier Matilda and Valentine heavy tanks had been replaced by the Churchill, which was no faster but was very heavily armoured (up to 152 mm). In its early versions it also was armed with 2 pdr and 6 pdr weapons, but by 1944 it carried a 75 mm gun.

This weapon was also mounted in the Cromwell Cruiser Tank which replaced the Crusader. It still had a respectable turn of speed but thicker armour and was altogether a more viable fighting machine.

The M4 Sherman had itself gone through various modifications and, by 1944, existed in a bewildering variety of types with different suspensions, engine arrangements, crews and guns; the best of all these was the 17 pdr-armed Firefly operated by the British.

The Russians had not been idle either. Later versions of the KV-1 were fitted with a new turret mounting an 85 mm gun and were thus designated KV-85. The T-34 was later similarly up-gunned and designated T-34/85. The KV-85 itself became the forerunner of a new class of heavy tanks, the IS-1, -2 and -3 which featured progressively thicker and more acutely sloped armour plate and a powerful 122 mm gun.

In Germany the trend was also towards ever heavier tanks, resulting in the Tiger II or King Tiger which resembled the Panther in appearance but was larger, heavier, better armoured and mounted a long-barrelled high-velocity version of the earlier 8.8 cm gun fitted in the Tiger I. In common with the Tiger I and Panther, however, it was severely underpowered and could no better stand up to attacks by British rocket-firing Typhoon aircraft than any other German tank.

What was more significant than the new designs, though, was the quantities in which they were produced, since by 1944 American and Soviet arsenals were turning out tanks in their hundreds practically daily, while German industry had been so pummelled by Allied bombing raids that their tanks were only coming off the production lines in tens.

Victory thus seemed to be in the air until Operation Market Garden, Montgomery's ambitious plan to breach the River Rhine by airborne assault. But the tragedy of Arnhem and the subsequent German winter 1944 Ardennes offensive—the 'battle of the bulge'—were the last shots in Hitler's locker. Besieged from all sides, the Führer committed suicide and, on May 7 1945, Grand Admiral Doenitz signed the document of surrender. The war in the Far East would continue until August 14, but the end was in sight.

It will be apparent from the foregoing that the Second World War was totally unlike anything previous. It was truly a global conflict and for that, amongst other, reasons, forms a particular problem and challenge for the wargamer. For that reason, the remarks which follow are more detailed than in previous chapters.

Tank battles in miniature

Up until 1939, warfare took place on a more or less individual basis. Units, whether cohorts and legions or battalions and brigades, can be formed on a scaled-down basis and brought into conflict to provide an afternoon's or weekend's pleasure for the players. Even the trench warfare of 1915–18 can be reduced to manipulative size. When one comes to consider the range, speed and firepower of Second World War tanks, however, it becomes apparent that different parameters must be applied. In the Napoleonic period, a divisional or even Corps level encounter can be simulated with little loss of realism; to attempt the same for the Second World War without a week's time and a cricket ground for the 'table' would be a hopeless exercise.

For the first time, here one has to consider a different scale to that used throughout this book. It is one thing to have an infantryman marching at, say, 100 yards per minute; it is quite another to contemplate a tank travelling at 25 mph (733 yards per minute) with a weapon range conceivably measured in miles.

The whole situation becomes even more convoluted if you try to incorporate all the diverse elements of Second World War combat, from Molotov Cocktails to strategic bombers or bayonets to battleships. For this reason, those who fight Second World War engagements tend to separate their activities, and there are ample sets of playing rules available on the market to satisfy all tastes, whether you wish to fight individual platoon-level infantry battles, encounters between U-boats and convoys, bomber sorties over London or Berlin, or even atomic strikes upon Japan.

The most common form of Second World War encounter, however, is the

Armoured warfare 1939-45

Italy 1944. German Panthers behind a 'barbed wire' barricade should make mincemeat of this Anglo-American Sherman and Grant despite their supporting infantry in the M3 half-tracks.

straightforward tank versus tank battle, and for those who wish to delve into the subject more deeply there are four books available in Patrick Stephens' *Tank Battles in Miniature* series, which cover respectively the Western Desert, Russian, North-West European and Mediterranean campaigns. Each of these books includes a more detailed campaign account than it is possible to give here, together with detailed technical data and chapters on such subjects as communications, lines of supply, navigation, minefields, artillery, infantry weapons and many other subjects.

They all have one thing in common in that they depart from the 25 mm (or 15 mm) figures used in other historical periods and go right down to tiny 1:285 or 1:300 scale tanks, which are commercially available in white metal castings at approximately the same price as 25 mm infantry figures. This small scale, in which a heavy tank is little more than half an inch long, permits the use of much smaller movement and weapon range scales in some form of proportion. For example, a metre on the wargames table may be deemed to represent a kilometre of real ground, in which case a tank moving at 30 km/h would cover 50 cm in a one minute game move. This is rather excessive and, as a result, half a minute or even less is taken as the average game move time.

The playing rules which follow this chapter are, for space reasons, necessarily devoted purely to tank versus tank combat during World War 2. The above-mentioned books will give further ideas to those who wish to bring in infantry or other forms of weaponry; there is just insufficient room here. The rules are designed to produce fast-moving and exciting games whose elements can be learned very quickly—but, as in chess, the gods of victory will favour the thoughtful rather than the impulsive player.

Rules for armoured wargaming

Once again, we have to make a break with the preceding rules. It is one thing to pitch a Zulu against a British redcoat, or Bulldog Drummond against the young Erwin Rommel, and quite another to match a PzKpfw IVF2 against a T-34. And just as warfare itself becomes more complex with technological innovation, so do wargames rules which attempt to cover the whole gamut of a period. So, just as I have been forced through lack of space in the preceding chapter to limit my remarks almost exclusively to armoured warfare, so the following rules are designed for pure tank versus tank encounters—the most popular aspect of World War 2 wargaming by any yardstick.

Whereas previously a 25 mm figure scale has been the 'norm', here we need an entirely new scale as a result of the movement rates and weapon ranges being so extreme. Many people, I must point out, *do* fight World War 2 battles with the Airfix-type model tanks and figures; and by request I once wrote a small book on it; but it is rather unsatisfactory to have two model tanks, each around three inches in length, facing each other over a table five feet wide, and trying to pretend that the intervening distance represents several hundred or thousand yards. Yet board games of the Avalon Hill or SPI variety lack the *visual* appeal of the 'figure' game. Is there a compromise? Fortunately, yes.

The scale most commonly adopted for 20th century armoured warfare is 1:285 or 1:300, the two being aesthetically close enough that the difference does not matter. In this scale, the average model tank, cast in white metal alloy, is only around a quarter to half an inch in length, yet can be almost perfectly proportioned and still contain a high degree of detail, even down to some models having revolving turrets! This still does not completely remove the ground scale problem, as at 1:300 a table five feet wide still only represents roughly a third of a mile. However, in the same way that one can compromise and make a 25 mm figure shuffle forward a mere six inches (representing 60 paces or roughly 50 yards) a minute in the earlier rules, even though the ground and figure scales are totally unrelated, we can do the same with these tiny tanks and still achieve battles which are visually appealing.

The basic ground rules for this type of wargame are essentially the same as what has gone before. All movement is simultaneous; each model tank represents one real tank; the game move represents 30 seconds of actual battle time; and the ground scale is 50 yards to the inch. On this basis, a tank moving at an average 25 mph can traverse $\frac{25 \times 1760}{120} \div 50$ inches per move, or just over seven inches, this representing the 366 yards at 50:1 that a real vehicle travelling at 25 mph would cover in 30 seconds. When you come on to more sophisticated sets of tank battle rules, you will find a wide variety of ground and movement scales, but this will give you the feel of things.

You will also find that most rules give separate factors for every individual type of vehicle. That is not practicable here for space reasons so the tanks are categorised as follows: heavy tanks move at four inches per move; medium at eight; and light at 12. This *does* ignore fine distinctions which you will undoubtedly encounter later, as for movement purposes several medium tanks should count as light, and so on. However, let it stand for the time being.

For weapon purposes, 'heavy' tanks are classified as those mounting 88 mm guns and up; 'medium' between 50 mm and 76 mm; and 'light' anything below 50 mm.

For armour purposes, 'heavy' tanks are classified as those with 90 mm or more of frontal armour, 'medium' as those with between 40 and 90 mm, and 'light' as those with below 40 mm.

A 'heavy' gun is capable of penetrating 'light' armour at any range up to 2,000 yards (40 inches), 'medium' at any up to 1,200 yards (24 inches) and 'heavy' at any up to 500 yards (ten inches); a 'medium' gun is capable of penetrating 'light' armour at up to 1,600 yards (32 inches), 'medium' at up to 900 yards (18 inches) and 'heavy' at only 300 yards (six inches); 'light' guns are capable of penetrating 'light' armour at up to 600 yards (12 inches), 'medium' at up to 200 yards (four inches), and 'heavy' not at all.

However, tanks generally had thinner armour at sides and rear than at the front, so when we came to the following firing rules we add +1 to all weapons when firing at a tank's side and +2 against a rear.

Once again, a tank may fire *or* move, but may not fire on the move. (This is over-simplifying because it *did* happen, but the reduced accuracy before the introduction of gyro-stabilisers and modern rangefinders made it a very chancy business.)

The procedure is as follows: first declare your intention of firing, and the target, *then* measure the range. Measuring beforehand is definitely cheating, but you soon develop an 'eye' for the distances your own guns are capable of achieving! If you are out of range, or the type of gun firing is incapable of penetrating the target's armour at that particular range, there is no effect.

If you *can* score a hit with a gun which is capable of penetrating the appropriate armour class (eg, medium gun against heavy armour at five inches), throw an ordinary dice; at long range a 5 or a 6 is a hit, at medium a 3, 4, 5 or 6, and at short anything except a 1. If the target is moving or behind cover, deduct −1 from your dice throw.

Having (hopefully) hit your target, you now assess damage by throwing an ordinary dice numbered one to six: 1 = target immobilised for one move but may still fire; 2 = target immobilised for three moves and may not fire for one move; 3 = target permanently immobilised and may not fire for three moves; 4 = target may still move but may not fire for one move; 5 = target may still move but may not fire for three moves; 6 = target totally destroyed.

These scores are accumulative in that, if you hit the same tank twice or more in succession, you inflict the next band of damage upon it. Thus, for example, if you scored a 3 on your first shot, permanently immobilising it, a second shot will automatically prevent it firing for a *fourth* move, and so on. The simplest way of recording the status of any tank at any given time without littering the table with markers is to paint a prominent number on its rear engine plate and record these down the side of a sheet of paper, with the status in any one move pencilled alongside.

The foregoing obviously simplifies the complexity of armoured warfare a great deal, to use a traditional English understatement, but it will produce a fast-moving and enjoyable game whose end results will not vary widely from those produced under far more sophisticated systems.

Appendices

Bibliography

The following is a *very* selective list, for reasons of space. However, most books included carry their own bibliographies which will extend the scope of your reading in the right directions.

Ancients

Armies and Enemies of Ancient China, by J. Greer, *Armies and Enemies of Ancient Egypt and Assyria,* by A. Buttery, *Armies and Enemies of Imperial Rome,* by P. Barker, *Armies of the Greek and Persian Wars,* by R. Nelson, *Armies of the Macedonian and Punic Wars,* by P. Barker, *Setting up a wargames campaign,* by T. Bath and *Warfleets of Antiquity,* by R. Nelson, all published by the Wargames Research Group, 75 Ardingly Drive, Goring by Sea, Sussex, in addition to their excellent sets of playing rules for all periods. Also *Alexander the Great's Campaigns,* by P. Barker, from Patrick Stephens Ltd.

Mediaeval

Armies of the Dark Ages, Armies of Feudal Europe and *Armies and Enemies of the Crusades,* all by I. Heath and published by the Wargames Research Group at the above address. Also *A Wargamers' Guide to the Crusades,* by the same author, published by Patrick Stephens Ltd. However, the classic work on this period remains Sir Charles Oman's *A History of the Art of War in the Middle Ages,* recently republished by Methuen.

16th and 17th centuries

Renaissance Armies, by G. Gush, also published by Patrick Stephens Ltd; *With Pike and Shot,* by C. Wesencraft, published by The Elmfield Press, and *Wargaming Pike and Shot,* by D. Featherstone, are good wargaming guides.

18th century

The Art of Warfare in the Age of Marlborough, by D. Chandler, published by Batsford, and *Arms and Uniforms: The Lace Wars* Volumes 1 and 2, by L. and F. Funcken, published by Ward Lock, provide a good introduction.

Napoleonic

Napoleon's Campaigns in Miniature, by B. Quarrie, also published by Patrick Stephens Ltd; *The Campaigns of Napoleon,* by D. Chandler, published by Weidenfeld & Nicholson; *Arms and Uniforms: The First Empire,* by L. and F. Funcken, published by Ward Lock, plus many more.

19th Century

The Penguin Book of the American Civil War, by B. Catton, *The Franco-Prussian War,* by R. May and G.A. Embleton, published by Almark and *Victorian Military Campaigns,* edited by B. Bond, published by Hutchinson, are good introductions to the varying aspects of this period.

World War 1

The British Army 1914-1918 and *The German Army 1914-1918,* by D. Fosten and R. Marrion, both published by Osprey.

World War 2

The *Tank Battles in Miniature* series, by D. Featherstone and B. Quarrie, published by Patrick Stephens Ltd, is an ideal introduction to the armoured warfare of this period, although some titles are now out of print so you may have to hunt for them.

General

D. Featherstone's *Wargames Through the Ages* series, published by Stanley Paul, as well as his earlier *War Games, War Games Campaigns* and *Advanced War Games,* all provide stimulating reading. D. Nash's *Wargames* from Hamlyn is a colourful paperback on the subject, J. Tunstil's *Discovering Wargames* from Shire Publications is also a stimulating booklet, and T. Bath's *Setting up a wargames campaign* from WRG is another excellent purchase.

For those wishing to explore other avenues, Patrick Stephens Ltd also publish *Air Battles in Miniature,* by M. Spick, and *Sea Battles in Miniature,* by P. Hague.

Model figure manufacturers and suppliers

Editor's note: the following names and addresses are believed to be correct at the time of writing but are subject to change without notice during the currency of this volume. In order to ascertain the precise ranges covered by each manufacturer, the best solution is a stamped, self-addressed envelope requesting a current price list to the addresses below or those published in advertisements in the various military, modelling and wargaming magazines.

Neither the contributors, editor nor publishers of this book can accept responsibility for errors in the following list, although all names and addresses have been checked conscientiously. Most firms operate a mail order service but, again, the contributors, editor and publishers of this volume can accept no responsibility for disputes which may arise between readers and any firms from whom they have ordered goods.

We apologise to any manufacturer of wargames goods who may have been inadvertently omitted from this list and request that they notify us so that, hopefully, additions and corrections, if any, may be incorporated in any future revised edition(s) of this book.

Advance Guard, 114 Crawford Street, Motherwell, Lanarkshire ML1 3BN. 15 mm Napoleonics and others.

Airfix Products Ltd, Haldane Place, Garratt Lane, London SW18 4NB. Large range of 20 mm plastic figures and 1:76 scale military vehicles and equipment. (Do not supply direct to the public.)

Asgard Miniatures, 14 Commerce Square, High Pavement, Nottingham. 25mm Ancients and Mediaevals.
Castile Miniatures, 20 Rankin Street, Carluke, Lanarkshire. 25 mm Ancients and Mediaevals.
Citadel Miniatures Ltd, Newark Folk Museum, 48 Millgate, Newark, Notts. Mostly fantasy figures, but some Ancients and Mediaevals.
Dixon Miniatures, Ash Grove, Royles Head Lane, Longwood, Huddersfield HD3 4TU. 25 mm Ancients and Mediaevals.
Freikorps 15, 30 Cromwell Road, Belfast 7, Northern Ireland. Wide range of 15 mm figures in various periods.
Greenwood & Ball (Garrison), Unit 27, Bon Lea Trading Estate, Thornaby, Co Cleveland. Wide range of 25 mm figures in a variety of periods.
Heroics and Ros Figures, PO Box 26, Rectory Road, Beckenham, Kent BR3 1HA. Large range of 1:300 scale tanks and figures.
Hinchliffe Models Ltd, Meltham, Huddersfield HD7 3NX. Very large range of 25 mm figures for most periods plus 20 mm World War 2 figures and guns.
Jacklex, The Model Shop, 190-194 Station Road, Harrow, Middx. Include World War 1 figures in their 20 mm range.
Jacobite Miniatures, 543 Gorgie Road, Edinburgh EH11 3AR. Good range of 15 mm figures.
Peter Laing, 'Minden', Sutton St Nicholas, Hereford. Probably the largest selection of 15 mm figures in various periods.
Lamming Miniatures, 254 Wincolmlee, Hull HU2 0PZ. Good range of 25 mm figures in several periods.
Matchbox, Lesney Products Ltd, Lee Conservancy Road, London E9 5PA. Remarks as for Airfix.
Micro-Mold Products, Station Road, East Preston, West Sussex BN16 3AG. Vacuum-formed wargames buildings and accessories in 1:300 and 25 mm.
Mike's Models, 38 Queens Road, Brighton BN4 4RQ. Large 15 mm range of Ancients and Mediaevals.
Miniature Figurines Ltd, 1-5 Graham Road, Southampton SO2 0AX. Probably the world's largest manufacturer of 25 mm figures in virtually every period.
New Hope Design, Rothbury, Northumberland NE65 7QJ. Minutely detailed 1:285 scale tanks and military equipment.
Rose Miniatures, 15 Llanover Road, London SE18. Good range of 20/25 mm figures.
Skytrex Ltd, 28 Brook Street, Wymeswold, Leics. Good range of 1:300 scale tanks, also books, rules, dice and other accessories.
Spencer Smith, 66 Long Meadow, Frimley, Camberley, Surrey. 30 mm American War of Independence and ACW figures.
Tradition, 5a & 5b Shepherd Street, Mayfair, London W1. Good variety of 25 mm figures.
Warrior, 44 Candleriggs, Glasgow. 15 mm and 25 mm figures in several periods.

Few model shops stock more than one or two ranges, and it is almost inevitable that they will not have the particular figures *you* want in stock (although most will order for you), so mail order is the best way of obtaining wargames figures. One word of warning, however: if you send cash with your order, *do* make sure you mail it by Registered Post.